Current Topics in Plant Physiology:
An American Society of Plant Physiologists Series
Volume 1

Plant Reproduction:
From Floral Induction to Pollination

Edited by
Elizabeth Lord
George Bernier

Proceedings
12th Annual Riverside
Symposium in Plant Physiology
January 12-14, 1989

Department of Botany and Plant Sciences
University of California, Riverside

Published by:
American Society of Plant Physiologists
15501-A Monona Drive
Rockville, Maryland 20855

Copyright 1989. All rights reserved. No part of this publication may be reproduced without the prior written permission of the publisher.

Library of Congress Cataloging in Publication Data

Main entry under title:

    Plant Reproduction: From Floral Induction to Pollination

Current Topics in Plant Physiology: An American Society of Plant Physiologists Series, Volume 1.
Includes bibliographies and index.

    1. Plant-Reproduction--Congresses. 2. Floral Induction--Congresses. 3. Pollination--Congresses.

    I. Lord, Elizabeth M., 1949- . II. Bernier, Georges, 1934- . III. University of California, Riverside. Dept. of Botany and Plant Sciences. V. Title.

Library of Congress Catalog Card Number: 89-80454
ISBN 0943088-14-3

Standing (Left to Right):
E. Meyerowitz, J. M. Kinet, R. Pharis, I. Murfet, S. McCormick, J. Deitzer, H. Dickinson, P. Green, J. Mascarenhas
Sitting (Left to Right):
R. Meeks-Wagner, C. McDaniel, V. Sawhney, G. Bernier, E. Lord, S. O'Neill, M. Anderson, I. Sussex

Standing (Left to Right)

J. Medford, S. Smith, A. Orr, K. Sangrey, M. Williams, S. Singer, D. Jegla, P. Lumsden, P. Bedinger, K. Eckard, P. Saradhi, J. Cross, P. Becraft, N. Lehmann, K. Feldmann

Sitting (Left to Right)

J. Hill, _____, L. Sanders, W. Crone, S. Brown, T. Sims, P. Schultz, B. Veit, P. Eastman

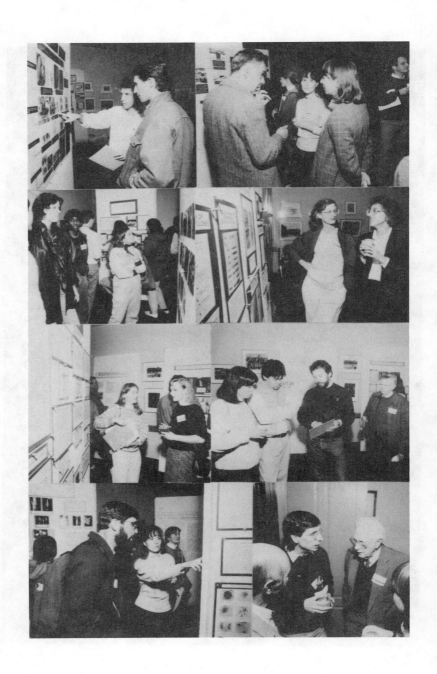

CONTRIBUTORS

G. Aivalakis
Eduardo Almeida
Marilyn A. Anderson
Mitchell Altschuler
David M. Bashe
Carole L. Bassett
Phil Becraft
Patricia Bedinger
Arnold J. Bendich
Georges Bernier
S. K. Bhadule
S. R. Bowley
John L. Bowman
Carl J. Braun
Sherri M. Brown
Reed Clark
Adrienne E. Clarke
A. W. Coleman
Pamela Collins
J. L. Corriveau
F. Cremer
W. Crone
John W. Cross
Martha L. Crouch
Gerald F. Deitzer
E. S. Dennis
Hugh G. Dickinson
J. Dommes
P. A. Eastman
Paul R. Ebert
K. J. Eckard
Michael D. Edgerton
J. Scott Elemer
Carole J. Elleman
B. E. Ellis
Norman Ellstrand
Lloyd T. Evans
K. A. Feldmann

Teresa S. Findlay
Michael Freeling
L. J. Goff
Robert B. Goldberg
K. S. Gould
Paul B. Green
Teresa Gruber
Sarah Hake
Douglas A. Hamilton
S. Han
Cole H. Hannon
Nic Harberd
Clare A. Hasenkampf
J. P. Hill
Sarah C. Huber
Martin J. G. Hughes
Dorothy E. Jegla
Brian R. Jordan
A. Kelly
Jean-Marie Kinet
Roderick W. King
Harry Klee
Jeffrey Labovitz
Susan Larabell
Naida Lehmann
Charles S. Levings, III
Eliezer Lifschitz
Elizabeth M. Lord
Peter J. Lumsden
M. Lund
Hong Ma
Lewis N. Mander
Joseph P. Mascarenhas
Sheila McCormick
Carl N. McDaniel
June Medford
D. Ry Meeks-Wagner
Bastiaan J. D. Meeuse

Elliot M. Meyerowitz
Debra Mohnen
Ian C. Murfet
John Nason
June Nasrallah
Mikhail Nasrallah
A. Neal
John Okuley
Sharman D. O'Neill
Alan R. Orr
Reid G. Palmer
Thomas Patterson
W. J. Peacock
Richard P. Pharis
R. Scott Poethig
P. L. Polowick
H. Y. Mohan Ram
Ilya Raskin
Nicolas Rasmussen
R. Rastogi
M. S. Reid
C. Daniel Riggs
L. C. Sanders
Karla A. Sangrey
P. Pardha Saradhi
R. H. Sarker

Rolf Sattler
Vipen K. Sawhney
Patricia J. Schulz
S. Shannon
Thomas Sims
Susan R. Singer
Halina Skorupska
Hanna Skubatch
Helen Slade
S. M. E. Smith
A. D. Stead
Herbert Stern
Ian Sussex
David Twell
Virginia Ursin
C. Van de Walle
Bruce Veit
Usha Vijayraghavan
J. Wahleithner
Thomas Whelan
R. White
Michelle H. Williams
Rod Wing
Judy Yamaguchi
Marty Yanofsky
Fan Zhang

EDITORS' INTRODUCTION

The products of plant sexual reproduction account for nearly 80% of the world's edible dry weight, but the processes leading to flower, fruit, and seed formation remain poorly known. After more than a decade of neglect, there is currently a mounting interest in the understanding of these processes. This arises from the recent introduction of novel conceptual approaches as well as new technologies making significant advances now possible.

Part of the past neglect is probably attributable to the almost incredible complexity of plant reproduction. In contrast to animals, plants cannot use a preexisting system of organs for reproduction. From the very beginning, they need to develop all parts of their reproductive apparatus and, for polycarpic species, need to repeat that each time they reinitiate sexual reproduction. The processes of gamete and zygote formation thus occur in step with flower morphogenesis. In fact, the manufactured products and the factory are made in a closely coordinated fashion and, once the products are made, the useless parts of the factory are rapidly destroyed.

Despite the fact that sexual reproduction is a unitary and basically continuous process, it is usually divided for convenience into successive steps. These may not react similarly to environmental factors and, since they are differently controlled by internal agents, they have been investigated separately. This is the basis for having in this symposium this unique assemblage of topics which spans floral induction, evocation, flower development, and pollination.

The simplest systems of making flowers, pollen, and seeds are the tobacco thin cell layers of Tran Thanh Van. Inside these tiny explants there is, however, a complex array of interactions which determine the kind of morphogenetic program expressed, and the locations of flower initiation are set, or at least anticipated, by the formation of tracheary centers. Thus, escape from inter-tissue or inter-organ correlations that were often considered to hinder work with intact plants is not, and cannot presumably be, achieved totally by extreme simplification of the experimental systems. This leads us to consider that the steps of the reproduction process are integrated systems of changes occurring at several levels of organization, from molecular to morphological. Understanding these systems will require a combination of classical and molecular genetics, cell biology, physiology, and developmental morphology. Thus, our aim in organizing this meeting was to bring together experts in these diverse disciplines and try to achieve an up-to-date synthesis of their complementary approaches.

Such a synthesis will not be easy to reach in the short term, essentially because we are all working on different plant species. During the symposium, all speakers have used a case-history approach, and there were thus talks on pea, tobacco, tomato, corn, rapeseed, and other *Brassicas*, *Lolium*, *Pharbitis*, *Arabidopsis*, *Sinapis*, *Lilium*, *Tradescantia*, etc., and if more than one presentation was made on a given plant, this was just by chance! The reasons of this situation are multiple. Pea, corn, and tomato rank among the genetically best-known of all plants, and *Arabidopsis*, with its small genome and many reproductive mutants, is best suited for a combination of molecular and genetic approaches. Tobacco is ideal for tissue culture. In studies where population synchronization is of prime importance, plants like *Pharbitis*, *Sinapis*, or *Lolium* are selected because they can be induced to flower after exposure to a single photoinductive cycle. For surface growth analyses, the large flower of *Lilium* is ideal.

In such a situation, at least part of the results are due to the specific behavior of the investigated species or variety. This may create some temporary confusion which will only be removed by comparing various genotypes of a given species and several unrelated species. Also, the reaching of generalizations about plant reproduction will necessitate that the successive steps of the process be studied in the same plant(s) in order to understand how these steps are interrelated to each other.

We all know that plants have to face, in their daily life, environments which are often unpredictable (even if these environments are relatively stable when considered over longer periods). They have adopted plastic developmental strategies to meet these challenges. Floral induction may be completed by most plants in several alternative sets of environmental conditions. Similarly, shoot meristems can apparently use alternative sets of cellular events in different conditions to achieve floral evocation. In various temperature regimes, the growth parameters of all primordia of the floral organs in *Silene* are variously altered, yet the relative organ proportions and, thus, the final form of the flower, remain constant. The basic structure of the embryo sac is conserved, but the ways used to develop this structure are multiple. Finally, flowering plants often combine simultaneously two or more reproductive strategies -- from sexual to asexual reproduction via apospory, diplospory, etc. -- or can switch between different strategies in their lifetime. Perhaps the finest example of this is the ontogenetic shift from closed (cleistogamous) self-pollinated flowers to open cross-pollinated flowers and back to cleistogamy in the developing inflorescences of a number of cleistogamous species..

Not all characteristics of plant development are plastic, however, and both plastic and stable aspects can exist side by side. Thus, a basic problem will be to determine what are the plastic and stable processes at each step of reproduction and at the various levels of organization. Particularly, it will be of paramount importance to know whether plants are using entirely, or only partially, different sets of genes in alternate developmental pathways.

The editors want to acknowledge Cindi McKernan of the Department of Botany and Plant Sciences at UCR for her expert advice in organizing the symposium and for her untiring attention to all the details that make such a gathering a success. We also appreciate the assistance of Robert Leonard, Chairman of the Department of Botany and Plant Sciences, for his financial assistance. Special thanks go to Aileen Wietstruk and Patti Fagan for their expert advice on editing, and for their skills in the production of the camera-ready copy for this volume. We thank Kathleen Eckard for her careful organization of the index. We would also like to acknowledge the graduate students, postdocs, and SRAs in the Department of Botany and Plant Sciences for their assistance with the symposium.

Finally, we sincerely thank the following sponsors for their financial support: USDA, Beckman Instruments, Inc., Monsanto Co., Pioneer Hi-Bred International Inc., and Zoecon Research Institute. We also gratefully acknowledge the support provided by Chancellor Rosemary S. J. Schraer, UCR; Seymour Van Gundy, Dean of the College of Natural and Agricultural Science, and the University of California Biotechnology Research and Education Program.

Georges Bernier
Elizabeth Lord

FOREWORD

With this volume, the American Society of Plant Physiologists continues its series of publications on timely topics in plant physiology. Publication of proceedings devoted to focus areas, such as the present one on plant reproduction from floral induction to pollination, is designed to share information from the symposia with other scientists. This and future books in the series will be identified as a volume in "Current Topics in Plant Physiology: An American Society of Plant Physiologists Series." It is the wish of the Publications Committee and the Executive Committee to make these publications as useful as possible. To this end, copies of this publication and publications from previous years are available at an affordable price from the American Society of Plant Physiologists, 15501 Monona Drive, Rockville, Maryland 20855, Telephone (301) 251-0560.

The ASPP Publications Committee
Jack C. Shannon, Chairman

Gerald E. Edwards	Chris R. Somerville
James E. Harper	Joseph E. Varner

ASPP symposia publications from previous years are:

1982: CRASSULACEAN ACID METABOLISM, Eds., I. P. Ting and M. Gibbs

1983: BIOSYNTHESIS AND FUNCTION OF PLANT LIPIDS, Eds., W. W. Thomson, J. B. Mudd and M. Gibbs

1984: STRUCTURE, FUNCTION AND BIOSYNTHESIS OF PLANT CELL WALLS,
  Eds., W. M. Dugger and S. Bartnicki-Garcia

1985: REGULATION OF CARBON PARTITIONING IN PHOTOSYNTHESIS TISSUE, Eds., R. L. Heath and J. Preiss

1985: INORGANIC CARBON UPTAKE BY AQUATIC PHOTOSYNTHETIC TISSUE,
  Eds., W. J. Lucas and J. A. Berry

1985: EXPLOITATION OF PHYSIOLOGICAL AND GENETIC VARIABILITY TO ENHANCE CROP PRODUCTIVITY,
  Eds., J. E. Harper, L. E. Schrader, and R. W. Howell

1986: MOLECULAR BIOLOGY OF SEED STORAGE PROTEINS AND LECTINS, Eds., L. M. Shannon
  and M. J. Chrispeels

1986: REGULATION OF CARBON AND NITROGEN REDUCTION AND UTILIZATION IN MAIZE,
  Eds., J. C. Shannon, D. P. Knievel and C. D. Boyer

1987: PLANT SENESCENCE: ITS BIOCHEMISTRY AND PHYSIOLOGY, Eds., W. W. Thomson,
  E. A. Nothnagel and R. C. Huffaker

1987: PHYSIOLOGY OF CELL EXPANSION DURING PLANT GROWTH, Eds., D. J. Cosgrove and D. P. Knievel

1988: PHYSIOLOGY AND BIOCHEMISTRY OF PLANT MICROBIAL INTERACTIONS,
  Eds., N. T. Keen, T. Kosuge, and L. L. Walling

1988: LIGHT-ENERGY TRANSDUCTION IN PHOTOSYNTHESIS: HIGHER PLANT AND BACTERIAL MODELS,
  Eds., S. E. Stevens, and D. A. Bryant

CONTENTS

Contributors vii

Editor's Introduction ix

Foreword xii

Contents xiii

List of Abbreviations xvi

Session I - Floral Induction

- Interaction Between Phytochrome and the Circadian Clock
  Mechanism to Control the Photoperiodic Induction of
  Flowering 1
  Gerald F. Deitzer

- Flowering Genes in *Pisum* 10
  Ian C. Murfet

- Molecular Analysis of Floral Induction in *Pharbitis nil* 19
  Sharman D. O'Neill

- Gibberellins and Flowering in Higher Plants: Differing Structures
  Yield Highly Specific Effects 29
  Richard P. Pharis, Lloyd T. Evans, Roderick W. King,
  and Lewis N. Mander

Session II - Floral Evocation and Initiation

- Events in the Floral Transition of Meristems 42
  Georges Bernier

- Floral Initiation as a Developmental Process 51
  Carl N. McDaniel

- Shoot Morphogenesis, Vegetative through Floral, from
  a Biophysical Perspective 58
  Paul B. Green

- Gene Expression During Floral Initiation...76
    D. Ry Meeks-Wagner, E. S. Dennis, A. Kelly, S. Shannon,
    R. White, J. Wahleithner, A. Neal, M. Lund, and W. J. Peacock

Session III - Flower Development

- Kinematic Analysis of Lily Flower Organs 82
    Elizabeth M. Lord and K. S. Gould

- Environmental and Chemical Controls of Flower Development..................95
    Jean-Marie Kinet

- A Molecular Approach to Flower Development
    in *Arabidopsis*...106
    Elliot M. Meyerowitz, John L. Bowman, Hong Ma,
    Usha Vijayraghavan, and Marty Yanofsky

Session IV - Pollen Development and Male Sterility

- Genes Expressed During Pollen Development..108
    Joseph P. Mascarenhas, Douglas A. Hamilton,
    and David M. Bashe

- Regulation and Development of Male Sterility in Tomato
    and Rapeseed...114
    Vipen Sawhney, S. K. Bhadule, P. L. Polowick and R. Rastogi

- Insights into the Texas Male Sterile Cytoplasm of Maize 121
    Charles S. Levings, III and Carl J. Braun

- Anther-Specific Genes: Molecular Characterization and
    Promoter Analysis in Transgenic Plants...128
    Sheila McCormick, David Twell, Rod Wing, Virginia Ursin,
    Judy Yamaguchi, and Susan Larabell

Session V - Pollination Biology/Incompatibility

- Molecular Physiology of the Pollen-Stigma Interaction
    in *Brassica*..136
    Carole J. Elleman, R. H. Sarker, G. Aivalakis, Helen Slade,
    and Hugh G. Dickinson

- The Genetics of Self-Incompatibility Reactions in *Brassica* and
    the Effects of Suppressor Genes 146
    Mikhail Nasrallah

- Molecular Genetics of Self-Incompatibility in *Brassica* 156
    June Nasrallah

- Molecular Genetics of Self-Incompatibility in
    Flowering Plants 165
    Marilyn Anderson, Paul R. Ebert, Mitchell Altschuler,
    and Adrienne E. Clarke

Poster Abstracts 174

Index 201

ABBREVIATIONS

2D	two-dimensional
ACC	1-aminocyclopropane-1-carboxylic acid
AVG	aminoethoxyvinyl glycine
BA	N6-benzylaminopurine
BC	blasticidin S
bp	base pair
CAB	chlorophyll a/b binding protein
CaMV	cauliflower mosaic virus
CCC	2-chloroethyltrimethyl ammonium chloride
CEPA	2-chloroethylphosphonic acid
CH	chasmogamous
CHA	chemical hybridizing agents
CK	cytokinin
CL	cleistogamous
CMS	cytoplasmic male sterility
D-day	first day of flowering
DF	daylight fluorescent
DG	diacylglyceride
DN	day neutral
DW	dry weight
F	fluorescent
FB	floral bud
FB7	day 7 floral bud
FLR	flower/leaf relativity
FR	far red light
FW	fresh weight
GAs	gibberellic acids
GC-SIM	gas chromatography-selected ion monitoring
GMS	genic male sterility
GP	glycoproteins
GUS	β-glucuronidase
IEF	isoelectric focusing
IgG	immunoglobulins
IMZ7	stem intermediate zone (non-inflorescence)
IP3	inositol-1,4,5-trisphosphate
KRI	Kovat's retention index
LAT	late anther tomato
lg1	liguleless1
li	invariant glycoprotein chain
LRGR	local relative growth rate
lw	lemon white
MAbs	monoclonal antibodies

MHC	major histocompatibility complex
mod1	modifier gene 1
Mp	miniplant
ms	male sterile
Pfr/Ptot	phytochrome photoequilibrium
PGR	plant growth regulator
pIs	isoelectric point
PI	pistillata
PIP2	phosphatidylinositol-4,5-bisphosphate
PR	pathogenesis related
R	red light
rbcS	small subunit of ribulose bisphosphate carboxylase
[rep]	*Nicotiana repanda* cytoplasm
RER	relative elemental rate of growth
RFLP	restriction fragment length polymorphism
Rts	retention times
SA	salicylic acid
SC	self-compatible
SI	self-incompatibility
S-gene	self-incompatibility gene
SLG	gene encoding the S-locus specific glycoprotein (SLSG) of *Brassica*
SLR1	S-locus related stigma specific gene
SLSG	S-locus specific glycoprotein
STS	silver thiosulfate
Su	suppressor gene
sup2	single recessive gene
SUP1	suppressor gene 1
SUP2	suppressor gene 2
TCL	thin cell layer
tmr	isopentenyl transferase locus from *A. tumefaciens*
tms2	gene for indol-3-acetamide amidohydrolase
ts	tasselseed
Tu	tunicate
[und]	*Nicotiana undulata* cytoplasm
VS	vegetative shoot
VS7	day 7 vegetative shoot
WL	white light
WT	wild type

INTERACTION BETWEEN PHYTOCHROME AND THE CIRCADIAN CLOCK MECHANISM TO CONTROL THE PHOTOPERIODIC INDUCTION OF FLOWERING

GERALD F. DEITZER

Department of Horticulture, University of Maryland, College Park, MD 20742, USA

Despite the fact that control of flowering by daylength has been intensively investigated for a long time (6), we still do not understand the fundamental cellular and biochemical basis for this regulation. Nevertheless, photoperiodically sensitive plants provide an important experimental tool that can be used to manipulate the onset of floral induction. The events that lead to floral induction take place in the leaves (15); however, the consequences of inductive treatments can only be assayed through measurements of the degree of floral transformation at apical and lateral meristems. Even in those instances where very early events in floral evocation, such as the enlargement of the apex or changes in carbohydrate and mitotic activity, are used to establish when this transition between vegetative and reproductive development occurs, they are still temporally and spatially separated from the inductive events in the leaves.

The involvement of phytochrome in the photoperiodic induction of flowering in leaves has also been known for some time (26). However, although phytochrome has been isolated, purified, cloned, and sequenced (11), we still do not know how it acts to mediate any of the numerous physiological responses that it controls. It has recently been found that phytochrome exists in at least two separate pools that differ kinetically, spectrally and antigenically (1). One of these pools predominates in etiolated tissue and the other in light-grown, green tissue. There is even some evidence from monoclonal antibody studies, that there is heterogeneity within the green tissue pool (1), suggesting the possibility that different phytochrome molecules may mediate different physiological responses.

There is also evidence that there are multiple modes of action of phytochrome involved in photoperiodic induction (31) that may be controlled by these different pools of phytochrome. One of these modes of action is related to the timekeeping aspect of photoperiodism that is initiated by a light/dark transition, and the other is a direct Pfr requiring response that leads to the production of a translocatable molecule that influences flowering at the apex. In this paper, we examine the dual nature of the phytochrome-mediated promotion of flowering in barley (*Hordeum vulgare* L.) and attempt to distinguish between the effect of far red (FR) light on the clock mechanism and the direct promotion of flowering. Our approach was to find a physiological response that was coupled to the same circadian clock mechanism as photoperiodic induction but had no direct relation with flowering.

METHODS

The flowering experiments described in this paper have been reported elsewhere (4, 5). In the gas exchange experiments, barley seeds were sown individually in 150-mm shell vials containing 0.8% agar in full strength Hoagland's solution. The vials were closed with a semipermeable cap to allow O_2 and CO_2 diffusion but retain H_2O to prevent drying. Prior to transfer to 12-h white light (WL) photoperiods, a hole was made in the cap to allow penetration of the coleoptile. This was sealed around the coleoptile with modeling clay. At the beginning of the 12-h photoperiods, four plants were placed in a six-chamber Plexiglas manifold that was designed to produce an air-tight seal around the caps of the vials and allow the whole seedling above the coleoptile to grow into the cylindrical chambers. The total volume of each chamber was 250 ml. These chambers were connected in parallel to an air supply that produced a constant CO_2 of 350 μl L^{-1} by mixing pure CO_2 through a micrometer valve with CO_2-free air that had been scrubbed through a series of soda lime columns. The CO_2 concentration was monitored continuously using an absolute IR red gas analyzer. The air stream was humidified by saturating the air at a dew point temperature calculated to produce 50% RH at 20°C and monitored continuously with an in-line miniaturized Visala humidity sensor. The flow rate was set to 3.0 L min^{-1} and monitored with a mass flow meter. The air was then divided into each of the six chambers, four containing seedlings and two empty to establish a zero baseline. The continuous air flow in each chamber was 0.5 L min^{-1} and each chamber was selected sequentially through activation of a solenoid by a stepping switch and routed to a differential IR red gas analyzer. The air from the first empty chamber served as a reference in all cases. After the gas was passed through the analyzer containing a water vapor detector, the water was reduced to a constant low level at 4°C before being passed through an analyzer containing a CO_2 detector.

The data from all instruments were monitored on a series of analog strip charts and then digitized and analyzed by computer. The data were averaged over the 15-min sample period and the results from all four plants were combined for each 15-min period and converted to instantaneous rates of net nmol s^{-1} plant^{-1} of CO_2 uptake or μmol s^{-1} plant^{-1} of H_2O vapor transpired.

The seedlings were monitored during the 4 d of 12-h WL photoperiods and the subsequent 3 d of WL, either with or without supplemental FR. At the end of the 3-d period, those seedlings in WL were transferred to darkness while those with supplemental FR were transferred back to WL for an additional 2 d. The lighting and other growth chamber conditions were the same as those used in the flowering experiments (4, 5).

RESULTS AND DISCUSSION

It is now firmly established in both SD plants (2) and LD plants (13) that the photoperiodic induction of flowering is based on an endogenous circadian rhythm. However, it has been much more difficult to demonstrate the rhythmic basis for induction in LD plants. This is due both to the fact that night breaks are much less effective and require much higher energies than those in SD plants and that extended dark periods are strongly inhibitory in LD plants. The ability of phytochrome to interact with this rhythm in LD plants has also been very difficult to establish. Vince-Prue (30) was able to demonstrate a time-dependent response to FR light when added to a daylength extension with dim red light in *Lolium temulentum*, and Deitzer *et al.* (4) have shown that there is a circadian rhythm in the sensitivity of *H. vulgare* to the addition of FR light during continuous WL. However, only one study (5) has reported an effect of light on both floral induction and on the timekeeping mechanism in LD plants. These results are summarized in Figure 1.

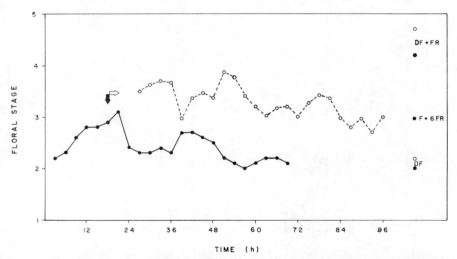

FIG. 1. Effect of 6 h of FR light on floral stage when added at various times from the onset of continuous WL (lower curve) and the effect of a second 6 h FR treatment added at various times following an initial treatment at the point marked by the arrows (upper curve). DF, daylight fluorescent; F, fluorescent. Redrawn from Deitzer *et al.* (5).

Flowering in barley is significantly enhanced when FR is added to 3 d of WL. When added for a 6-h period at various times during the 3-d WL period, the degree of promotion depends on when the FR is given. The sensitivity of plants to respond to FR light varies with circadian periodicity (lower curve, Fig 1). This change in sensitivity has been shown to be more than 50-fold greater at 18 h than at 6 h and the response is maximal between 710 and 730 nm (3).

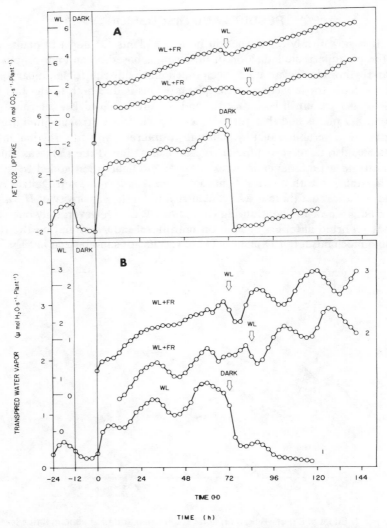

FIG. 2. Rates of net CO_2 uptake (A) and transpired H_2O vapor (B) monitored during the last day (-24 to 0) of the 4-d pretreatment, during 3 d of continuous WL (lower curves) or WL + FR (upper two curves) and for 2 d after return to either darkness or WL alone.

In addition to the promotion of flowering by a single 6-h FR treatment, there is a concomitant effect on the phase of this rhythm in sensitivity. This can be seen in the upper curve in Figure 1. Plants that had been given one 6-h FR treatment 18 h after the onset of the WL period, were given a second 6-h FR treatment at various times for an additional 3 d. As can be seen, there is an additional promotion of flowering caused by the second FR treatment, but the time of the next maximal response has shifted from 39 to 42 h to 33 to 36 h. Thus, the addition of FR light has two simultaneous effects on flowering in barley, it directly enhances flowering and it changes the phase of sensitivity to further enhancement by FR light. The former effect has been examined in some detail (3) and it has been concluded that it is mediated by phytochrome acting through what has been described as the high irradiance response (8). However, this is the first evidence for a phase shift caused by light in a LD plant. Very little is known about the mechanism for this response.

In order to examine the effect of FR light on the phase of the circadian rhythm, a much better defined physiological response that could be monitored in real time was sought. Based on experiments of Hillman (12) with *Lemna*, an attempt was made to establish a rhythm in gas exchange in barley leaves. The rates of both net CO_2 uptake and transpired H_2O vapor were monitored under the same conditions as those used for the flowering experiments. The results are reported in Figure 2A and Figure 2B. The three curves in each of the figures are essentially superimposable and so the scales have been offset for clarity. When the seedlings were transferred to continuous WL, there was little or no evidence for an overt rhythm in net CO_2 exchange (Fig. 2A, lower curve), nor was there any evidence for a rhythm in dark respiration when the lights were turned off. This is in marked contrast to the results of Hillman (12). There was also no effect of FR light on this pattern of CO_2 exchange (Fig. 2A, upper and middle curves).

Transpired H_2O vapor, on the other hand, shows a very pronounced circadian rhythm during WL (Fig. 2B, lower curve) but it damps out almost immediately upon return to darkness. Addition of FR for 6-h periods, as was used to demonstrate a rhythm in flowering, had no effect on this rhythm (unpublished results). However, when FR was added continuously from the beginning of the WL period (Fig. 2B, upper curve), the overt expression of the rhythm was abolished. The rhythm was reestablished when the FR was removed, and this rhythm had the same phase as that seen under WL without FR. When the onset of the FR was delayed by 12 h (Fig. 2B, middle curve) there was no apparent effect on the rhythm when compared to that in WL during the 3-d period. However, when the FR was removed, the phase of this rhythm was reset to the FR to WL transition and the next peak always occurred 15 to 18 h later. This results in a rhythm that is shifted by 12 h from that seen either under WL alone or that established after the removal of the FR that had been given from the start. Thus, FR affects the phase of the timing mechanism that regulates stomatal opening but does not affect net CO_2 exchange under the same conditions.

The effect of phytochrome on timekeeping has been studied most extensively in the SD plant, *Pharbitis nil*. In this very sensitive plant, flowering can be inhibited by a single brief pulse of red light (R) given during the course of an inductive dark period. A single 5-min pulse of R, in the presence of benzyladenine, is sufficient to initiate dark timing (25). A second R pulse, given at various times during an extended dark period, results in a circadian rhythm of floral inhibition. When WL is given to replace the R pulse and benzyladenine, the phase of this rhythm remains constant for durations as long as 6 h with the first peak of inhibition always occurring 15 h from the onset of the light period. With white light periods greater than 6 h, the first peak of inhibition always occurs 8 to 9 h from the end of the light period (17). Based on these results, it has been suggested that the rhythm of sensitivity to R, that is initiated by the onset of light, is suspended in continuous light longer than 6 h. When the continuous light is turned off, the rhythm restarts at a time set by the light/dark transition. This effect of light on the phase of the rhythm appears to be mediated by a different pool of phytochrome than that which causes the inhibition of flowering by a R pulse (18).

The effects of FR on both flowering and stomatal rhythms in barley show some similarities with the effects of R in *Pharbitis* except that, instead of rephasing to the dark/light transition, they rephase to a FR/WL transition. Thus, this effect of phytochrome appears to be mediated by a rapid change from a low to a high Pfr/Ptot (phytochrome photoequilibrium) since the phase shift occurs in constant light. The suppression of the stomatal rhythm in the presence of FR suggests that Vince-Prue's (31) interpretation of the phase response after continuous WL may be correct. However, the suspension of the rhythm is not essential to the phase response in barley since the phase is reset to exactly the same degree even when there is no effect on the overt expression of the rhythm (Fig. 2B, middle curve). It may be that the 6-h duration in WL in *Pharbitis* is required for the etiolated pool of phytochrome to decrease and the green tissue pool of phytochrome to establish regulation. This may explain why FR has an effect on the rhythm when given at the beginning of the WL period in barley but not after the plants have been in WL for 12 h. The concomitant nature of the effect of FR on floral promotion and the phase of the rhythm in barley also suggests that both responses may be mediated through the same mechanism, but more information is needed on the latter response to establish this.

Stomata function as multisensory turgor valves, opening and closing in response to a number of environmental stimuli by changing the turgor pressure in the guard cells surrounding the stomatal pore. This is accomplished through a chemiosmotic mechanism (32) by exchange of H^+ and K^+ ions, where the extrusion of H^+ ions is dependent on the operation of a plasma membrane ATPase pump. Stomatal rhythms have been reported in a number of species (22) and have been found to persist in constant light (9). The phase of these rhythms were found to be dependent on the length of the preceding dark period (19) and were reset to a light-off signal (20, 21). The phase of the stomatal rhythm reported here would appear to respond to a condition analogous to a light-on signal rather than a light-off signal.

The turgor changes that result in swelling and shrinking of pulvinar motor cells also involve massive movements of ions (27) which appear to be driven by proton extrusion (16) and regulated by a circadian rhythm. The phases of these rhythms have been shown to be regulated by phytochrome (28). While the molecular mechanism that regulates circadian rhythmicity remains unclear, a common mechanism appears to be ubiquitous in all eukaryotic organisms. This commonality has led to the recent suggestion (23) that it involves the turnover of membrane-localized phospholipids. According to this hypothesis, light interacts with some receptor at the plasma membrane to stimulate the hydrolysis of phosphatidylinositol-4,5-bisphosphate (PIP_2) by a phosphodiesterase (Phospholipase C) to diacylglyceride (DG) and inositol 1,4,5-trisphosphate (IP_3). The IP_3 mediates the release of calcium ions from stored intercellular reserves in the endoplasmic reticulum or the vacuole. The calcium binds to and activates a second messenger, calmodulin, which then activates a number of enzyme systems, especially protein kinases. Such kinases may ultimately lead to activation of gene expression and changes in metabolic function that are the realization of a light activated response. The IP_3 is rapidly hydrolyzed to inositol, which combines with DG through the action of a calcium-activated ATPase, thus completing the cycle.

While this pathway remains to be elucidated in higher plants, there is good evidence to suggest that phytochrome mediates calcium movement in cells (see 10 for a review). There is also some evidence to suggest that this constitutes the basic circadian oscillator (7) that may explain how light, as well as other environmental perturbations, interacts with the clock mechanism. In addition, light is known to regulate the transcription of a number of genes (for review, see 29) and this regulation is mediated by phytochrome. Kloppstech (14) was able to demonstrate that the steady state levels of mRNA encoding for the small subunit of ribulose bisphosphate carboxylase (rbcS) were regulated by a circadian rhythm. Recently, Nagy et al. (24) reported that there was a circadian rhythm in transcription of one member of the chlorophyll a/b binding protein (*Cab*) gene family and that phytochrome was responsible for this rhythmic expression. The circadian rhythm was maintained in transgenic tobacco suggesting that its expression is regulated at the transcriptional level. However, whether the clock mechanism, and the phytochrome interaction with it, occur directly at the gene level or indirectly through the mediation of a second messenger such as calmodulin is currently unknown.

LITERATURE CITED

1. CORDONNIER MM, H GREPPIN, LH PRATT 1986 Monoclonal antibodies with differing affinities to the red-absorbing and far red-absorbing forms of phytochrome. Biochem 25: 7657-7666
2. CUMMING BG, SB HENDRICKS, HA BORTHWICK 1965 Rhythmic flowering responses and phytochrome changes in a selection of *Chenopodium rubrum*. Can J Bot 43: 825-853
3. DEITZER GF 1983 Effect of far red energy on the photoperiodic control of flowering in Wintex barley (*Hordeum vulgare* L.). *In* W Meudt, ed, Strategies of Plant Reproduction. Allanheld Osmun, Totowa NJ, pp 99-116
4. DEITZER GF, R HAYES, M JABBEN 1979 Kinetics and time dependence of the effect of far red light on the photoperiodic induction of flowering in wintex barley. Plant Physiol 64: 1015-1021
5. DEITZER GF, R HAYES, M JABBEN 1982 Phase shift in the circadian rhythm of floral promotion by far red energy in *Hordeum vulgare* L. Plant Physiol 69: 597-601
6. GARNER WW, HA ALLARD 1920 Effect of the relative length of day and night and other factors of the environment on growth and reproduction in plants. J Agric Res 18: 553-606
7. GOTO K, DL LAVAL-MARTIN, LN EDMUNDS JR 1985 Biochemical modelling of an autonomously oscillating circadian clock in *Euglena*. Science 228: 1284-1288
8. HARTMANN KM 1966 A general hypothesis to interpret high energy phenomena of photomorphogenesis on the basis of phytochrome. Photochem Photobiol 5: 349-366
9. HEATH OVS, TA MANSFIELD 1962 A recording porometer with detachable cups operating on four separate leaves. Proc Roy Soc London Biol Sci 156: 1-13
10. HEPLER PK, RO WAYNE 1985 Calcium and plant development. Annu Rev Plant Physiol 36: 397-439
11. HERSHEY HP, RF BAKER, KB IDLER, JL LISSEMORE, PH QUAIL 1985 Analysis of cloned cDNA and genomic sequences for phytochrome: complete amino acid sequences for two gene products expressed in etiolated *Avena*. Nucl Acids Res 13: 8543-8559
12. HILLMAN WS 1976 Light timer interactions in photoperiodism and carbon dioxide output patterns. *In* H Smith, ed, Light and Plant Development. Butterworths, London, pp 383-397
13. HSU JCS, KC HAMNER 1967 Studies on the involvement of an endogenous rhythm in the photoperiodic response of *Hyoscyamus niger*. Plant Physiol 42: 725-730
14. KLOPPSTECH K 1985 Diurnal and circadian rhythmicity in the expression of light induced nuclear messenger RNAs. Planta 165: 502-506
15. KNOTT JE 1934 Effect of a localized photoperiod on spinach. Proc Am Soc Hort Sci 31: 152-154
16. LEE Y, RL SATTER 1987 H+ uptake and release during circadian rhythmic movements of Samanea motor organs: effects of mannitol, sorbitol and external pH. Plant Physiol 83: 856-862

17. LUMSDEN PJ, B THOMAS, D VINCE-PRUE 1982 Photoperiodic control of flowering in dark grown seedlings of *Pharbitis nil* Choisy. The effect of skeleton and continuous light photoperiods. Plant Physiol 70: 277-282
18. LUMSDEN PJ, D VINCE-PRUE, M FURUYA 1986 Phase shifting of the photoperiodic flowering response rhythm in *Pharbitis nil* by red light pulses. Physiol Plant 67: 604-607
19. MANSFIELD TA 1963 Length of night as a factor determining stomatal behavior in soybean. Physiol Plant 16: 523-527
20. MANSFIELD TA, OVS HEATH 1961 Photoperiodic effects on stomatal behavior in *Xanthium pennsylvanicum*. Nature 319: 974-975
21. MANSFIELD TA, OVS HEATH 1963 Studies in stomatal behavior IX Photoperiodic effects on rhythmic phenomena in *Xanthium pennsylvanicum*. J Exp Bot 14: 334-352
22. MEIDNER H, TA MANSFIELD 1965 Stomatal responses to illumination. Biol Rev 40: 483-509
23. MORSE MJ, RS CRAIN, RL SATTER 1987 Phosphatidylinositol cycle metabolites in *Samanea saman pulvini*. Plant Physiol 83: 640-644
24. NAGY F, SA KAY, N-H CHUA 1988 A circadian clock regulates transcription of the wheat *Cab*-1 gene. Genes & Dev 2: 376-382
25. OGAWA Y, RW KING 1979 Establishment of photoperiodic sensitivity bybenzyladenine and a brief red irradiation in dark grown seedlings of *Pharbitis nil* Chois. Plant Cell Physiol 20: 115-122
26. PARKER MW, SB HENDRICKS, HA BORTHWICK, NJ SCULLY 1946 Action spectrum for the photoperiodic control of floral initiation of short day plants. Bot Gaz 108: 1-26
27. SATTER RL, GT GEBALLE, PB APPLEWHITE, AW GALSTON 1974 Potassium flux and leaf movement in *Samanea saman*. I. Rhythmic movement. J Gen Physiol 64: 431-436
28. SIMON E, RL SATTER, AW GALSTON 1976 Circadian rhythmicity in excised *Samanea pulvini*. II. Resetting the clock by phytochrome conversion. Plant Physiol 58: 421-425
29. TOBIN EM, J SILVERTHORNE 1985 Light regulation of gene expression in higher plants. Annu Rev Plant Physiol 36: 569-593
30. VINCE-PRUE D 1965 The promoting effect of far red light on flowering in the long day plant *Lolium temulentum*. Physiol Plant 18: 474-482
31. VINCE-PRUE D, PJ LUMSDEN 1987 Inductive events in the leaves: time measurement and photoperception in the short day plant, *Pharbitis nil*. In JG Atherton, ed, Manipulation of Flowering. Butterworths, London, pp 255-268
32. ZEIGER E, AJ BLOOM, PK HEPLER 1978 Ion transport in stomatal guard cells: a chemiosmotic hypothesis. What's New in Plant Physiol 9: 29-32

FLOWERING GENES IN *PISUM*

IAN C. MURFET

*Department of Plant Science, University of Tasmania,
Hobart, Tasmania 7001, Australia*

Flowering and reproductive genes may be classified according to the developmental steps which they control. However, while each gene has a very specific action at the primary level, the consequences of a single gene change may be very widespread and diverse. For example, the *Sn* gene in *Pisum* confers a requirement for photoperiodic induction; delays flower initiation; retards flower bud, seed, and fruit development; extends the length of the reproductive phase; increases yield; and delays the onset of monocarpic senescence (1, 10, 11, 16, 18). Moreover, it also influences vegetative traits, *e.g.* branching habit (16). The primary action of a gene is often not immediately obvious and may require very detailed study for resolution. Suppose, for example, that genotype *A* responds to one inductive cycle while mutant *A'* requires several inductive cycles. This could mean that *A'* causes an inefficiency in the inductive process such that a lower level of flowering hormone reaches the apex after one inductive cycle (Fig. 1). The fault may be in the light perception system, in the synthesis process, or even in the transport mechanism from the leaves to the apex. Alternatively, mutant *A'* may have changed the sensitivity of the shoot apex such that a higher threshold level of the flowering signal is necessary to trigger flowering. Indeed, the alleles of the *Lf* locus in *Pisum* appear to act at the shoot apex in that manner (7). Gene *Lf* confers a higher threshold than *lf* and, with a similar age and background, *Lf* plants require more inductive cycles than *lf* plants (21). The recessive mutant, *veg*, also acts in the pea apex to block a step prior to floral morphogenesis, and *veg* plants are totally incapable of responding to the flowering signal (22).

The genes *lf* and *veg* are therefore evocation mutants, but they may also be viewed as sensitivity mutants. In contrast, the *Pisum* mutants *sn* and *dne* appear to operate in the leaves where they block, or partially block, steps in the synthesis of a graft-transmissible substance which functions as a flower inhibitor (7, 11, 14). Activity of the *Sn Dne* system is prevented by light (15), and the two dominant genes together confer a requirement for photoinduction. Thus, *sn* and *dne* may be considered induction mutants. However,

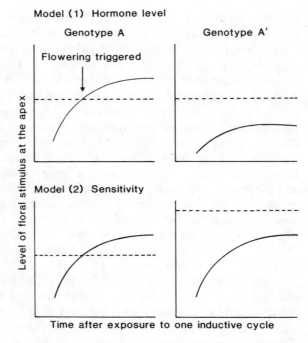

FIG. 1. Two models to explain a hypothetical case in which genotype A flowers in response to one inductive cycle, but mutant A' requires several cycles. Model (1) mutant A' reduces efficiency of induction in the leaves. Model (2) mutant A' reduces the sensitivity of the apex to the floral stimulus, i.e. the threshold level necessary to evoke flowering is raised.

they may also be viewed as synthesis mutants. Two further genes, *E* and *Hr*, influence the ontogenetic expression of the *Sn Dne* system. Gene *E* functions early in the life cycle to reduce activity of the *Sn Dne* system in the cotyledons (7). In contrast, gene *Hr* blocks the decline in *Sn Dne* activity which occurs with time in *hr* plants (8, 21).

Classification of genes according to whether they influence synthesis of, or sensitivity to, the controlling hormone has proved useful in the case of the internode length genes of *Pisum*, where, for example, a number of mutants are known to interfere with gibberellin (GA) synthesis or sensitivity to applied GA (19, 20). On the flowering side, we are handicapped by lack of a handle on the regulatory hormone(s). Nevertheless, the above six flowering genes do fit a synthesis-sensitivity scheme fairly well. The GA synthesis mutants in *Pisum* also have a side effect on flowering since the short internode, GA-deficient, types produce 10 to 25% fewer leaves in a given time than wildtype, tall counterparts (17). Hence, the mutants take longer to produce an open flower at any given node (5, 17). Further, the length mutant *lw*, which possesses normal levels of GA-like substances, causes up to a 25% increase in node of flower initiation (4).

NEW INFORMATION ON LOCI *SN*, *DNE*, AND *HR*

The action of genes at the *Pisum* loci *Lf*, *E*, *Sn Dne*, *Hr*, and *Veg* has been studied fairly extensively and the literature is covered in recent reviews

(11, 16). Even so, some genetic combinations have not been explored (with six loci and four alleles at *Lf*, 128 pure combinations are possible) and new information is still emerging. Segregation for the gene pairs *Sn-sn* and *Hr-hr* on background *lf e Dne* can readily be followed in SDs (8-12 h) by reference to node of flower initiation. The dihybrid ratio is 9 high node (*Sn Hr*):3 medium node (*Sn hr*):4 low node (*sn Hr* and *sn hr*) (8). With background *lf E Dne*, all plants flower at the low node. However, using an intermediate photoperiod (14 h) and two variables, number of reproductive nodes and distance (nodes) below the apical bud, the flower opens; it has now proved possible to obtain an equally clear 9:3:4 dihybrid ratio on an *lf E Dne* background (13). *Sn Hr* plants have a large number of reproductive nodes and the flowers open well below the apical bud, while genotypes *sn Hr* and *sn hr* have the smallest number of reproductive nodes and their flowers open closest to the apical bud. The *Sn hr* segregates occupy an intermediate, but discrete, position. It is worth noting that while SDs maximize activity of the *Sn Dne* system, an intermediate photoperiod optimizes simultaneous separation of these three groups.

Since the above variables effectively discriminate among genotypes differing at the *Sn* and *Hr* loci, it follows that they may be used to estimate the level of *Sn Dne* activity. One question of interest is whether the mutants *sn* and *dne* are leaky. Accordingly, the above variables were used to examine differences among plants belonging to the so-called early day-neutral class (11) and, in particular, to compare the genotypes *sn Dne* and *Sn dne* on an *Lf E hr* background. These two genotypes flower fairly early, about nodes 11 to 13, and the node of flower initiation is essentially unaffected by photoperiod (11). The results obtained in 8 h SD show that, compared with *sn Dne* plants, *Sn dne* plants have significantly more reproductive nodes and open their flowers further from the apical bud. However, the differences are quantitative and the distributions overlap. Nevertheless, the results indicate that mutant *dne* only partially blocks activity of the *Sn Dne* system. Previous work (1) shows that *Sn Dne* activity also retards pod growth. Again, in the present study, pod width 14 d after open flower was significantly less in *Sn dne* than *sn Dne* plants, further substantiating the leaky nature of mutant *dne*. However, the clearest evidence is coming from crosses involving an Lf^d background. The Lf^d allele imparts to the apex a high threshold, *i.e.* it requires a strong signal to flower. Conversely, we may view the apex as very sensitive to a flower inhibitor. Recombinant plants with genotype Lf^d *Sn Dne Hr* showed a massive response to photoperiod, with the flowering node rising from a mean of 23 in a 24-h photoperiod to a mean of 78 in an 8-h photoperiod. This behavior is very similar to that of genotype Lf^d *Sn Dne Hr* (9). Thus, on background Lf^d *Sn* gene, *Hr* is epistatic to the *Dne-dne* pair of alleles. In contrast, *sn* proved to be epistatic to the *Hr-hr* gene pair; *i.e.* the present results and those of Duchene (1) show that genotypes Lf^d *sn Dne hr* and Lf^d *sn Dne Hr* are both late-flowering, day-neutral types in which the flowering node is mostly around 21 to 24 even in 8 h SDs. Thus, the *sn* mutant, if not an amorph, certainly imposes a severe block to *Sn Dne* activity, while the *dne* gene is clearly leaky.

While the transition from the vegetative to flowering state is of central interest in any study of flowering, this step provides only a fleeting glimpse of hormone level as the threshold is intersected. However, where a gene product like that of the *Sn Dne* system influences several traits, it might be possible to use some of those traits to give a continuing estimate of gene activity (hormone level) within any one plant. Indeed, the distance the flower opens below the apical bud will possibly perform such a function since it declines with age in a continuous manner in many genotypes. Indeed, the later formed flower buds may open within the apical bud itself and, in some genotypes, particularly those with gene Lf^d, even the first-formed flowers may open within the apical bud (11). The index, flower/leaf relativity (FLR) (10, 11) covers both contingencies. If a flower opens before the subtending leaf is fully expanded, the FLR value is positive; otherwise, it is negative. For example, an FLR value of -2 means that there were two fully expanded leaves above the flower at the day the flower opened.

In genotypes such as *lf E Sn Dne hr*, the FLR value for each succeeding flower rises in a near linear manner throughout the reproductive phase. Another character, peduncle length, declines in a similar manner. The two traits show an inverse linear relationship for extended periods and, like FLR, peduncle length has proved very effective in discriminating between genotyes *SN* and *sn* in certain situations (11). Peduncle length is also a less difficult and more convenient character to score than FLR. The changes in FLR and peduncle length during ontogeny appear to reflect a decline in *Sn Dne* activity with age in *hr* plants. In plants with dominant *Hr*, *Sn Dne* activity persists for a prolonged period in SDs and peduncle length also shows little overall change, although some perturbations occur from time to time. Clearly, it would be useful to have some characters which reflected *Sn Dne* activity and changes in the level of flowering hormones prior to flower initiation. Branching pattern may prove useful in that respect.

NEW FLOWERING MUTANTS

Several additional flowering mutants have recently come under study at Hobart.

Suppressed Flower Bud Development. Gottschalk (2, 3) has obtained by mutation breeding certain lines, *e.g.* R20E, in which flower bud development is extensively suppressed in SD conditions. The flower initials do not develop beyond the stage of tiny buds which eventually wither and abort. The trait is also expressed, but to a lesser extent, in the LD phytotron conditions used by Gottschalk. Suppressed or retarded development of the first flower initials in SD conditions occurs, to a limited extent, in all genotypes with an early photoperiodic habit, *e.g. lf E Sn Dne hr* (7, 11, 18). However, our tests show that the failure of the flower initials to develop is certainly much more marked and extensive in R20E than in any of the normal early photoperiodic lines in our collection, *e.g.* lines 60 or 102. Crosses with R20E have raised certain questions which need to be resolved before the genetic basis for the behavior of R20E can be determined.

Mutant *dm* **(Diminutive)**. The recessive *dm* mutant arose from natural causes in Hobart line 89. The effects of *dm* on several traits are summarized in Table I. Flowering is delayed indefinitely in the background in which the mutant arose (Lf^d *sn Dne hr*). In the earlier background provided by gene *lf*, there is still a massive 2- to 3-fold increase in the node of flower initiation. As the name implies, internode length and leaflet size are markedly reduced. The leaflets are more ovate and darker green in color. The transition from two to more than two leaflets per leaf occurs much later in ontogeny than is normal and then only to a trifid form. The decrease in internode length is wholly due to a reduction in the number of cells per internode. There are floral abnormalities and the flowers are female sterile, but male fertile. This mutant warrants further study.

Mutant *det* **(Determinate)**. The recessive *det* mutant causes the shoot to terminate in a flower after a small number of normal axillary flowers have been produced (6, 23). This mutant is not to be confused with Mendel's fasciated type which is often erroneously described as having terminal flowers. The interaction with other flowering genes has recently been examined (12). The *det* mutant was received in line JI1358 which is very late flowering with a high response to photoperiod. A cross with one of our earliest, day-neutral lines (line 69, lf^a *E sn Dne hr*) produced an F_2 in which a wide range of flowering phenotypes occurred. A similar range of flowering

Table I. *Effect of the dm (diminutive) mutant on several traits relative to Dm counterparts*

Trait	Effect	Background genotype
Flowering	- delayed 2-fold	*lf E sn Dne hr*
	- delayed 3-fold	*lf E Sn Dne hr*
	- delayed indefinitely	Lf^d *sn Dne hr*
Internode length	- reduced by 50-65%	dwarf *le*
Leaf size	- reduced by 40-50%	"
Leaf shape	- more ovate	"
Leaf color	- darker green	"
Cells/internode	- reduced 65-75%	"
Cell length	- not reduced	"
Gynoecium	- carpels 1-3 (*Dm* = 1)	"
	- female sterile	"
Androecium	- stamens 8-12 (*Dm* = 10)	"
	- male fertile	"

nodes occurred in a 10-h photoperiod in both *Det* and *det* segregates, and there was nothing to suggest that the mutant *det* altered in any way the effect of the genes at the *Lf, Sn,* and *Hr* loci in node of flower initiation. However, *det* brought forward by several days (on average 4) the time of the first open flower in very early initiating photoperiodic segregates (genotype *lf^a E Sn Dne*). In these plants, termination of the shoot appeared to force a more precocious development of the flower buds into mature flowers. In another correlative phenomenon, *det* also strongly stimulated the outgrowth of laterals which, like the main shoot, eventually terminated in a flower. The gene *lf^a* did have a very marked effect on the expression of *det*. In cross 69 x 1358, *det* segregates never terminated without producing at least nine normal leaves. Plants with genotype *Lf det* commenced flowering at node 10 or above, and they produced no more than two axillary flowers before terminating. However, *lf^a det* segregates commenced flower initiation as early as node 5, and they produced up to six axillary flowers before terminating. Clearly, a minimum period of leaf formation must occur before expression of *det* can be triggered by the onset of flower initiation. The terminal flower is slightly offset from the vertical and, whether it is formed by direct conversion of the apical meristem itself or from a lateral flower primordium, remains to be determined.

Mutant gi (*Gigas*). Like the Maryland Mammoth strain of tobacco, a *gigas* pea mutant obtained by Dr. M. Vasileva (mutant III/83) towers above its normal counterparts in the field by reason of tardy flowering. The internode length of the *gigas* mutant is no greater than that of its progenitor, cv Virtus. The *gigas* mutant displays a quite extraordinary flowering behavior (Table II). Cv Virtus shows reactions typical of a late-flowering, quantitative LD line, and a genotype of *Lf Sn Dne hr* is assumed but not proved. The *gigas* mutant is very much later than its progenitor in 8 h, and only about half the plants had flowered after 261 d when the study in Table II was terminated (4-m plants are difficult to manage in a controlled environment facility!). With increasing photoperiod, the flowering node of the *gigas* plants declined to a minimum of 34 in 18 h, which was still twice that of its progenitor. However, in a 24-h photoperiod (8 h daylight + 16 h incandescent light at 3 ℓmol m^{-2} s^{-1}), the mutant plants remained vegetative indefinitely. When transferred from the 8-h to the 24-h photoperiod, the *gigas* plants flowered transiently at two to four nodes before reverting to the vegetative state. In the 24-h conditions, the *gigas* plants closely resemble the *veg* mutant. However, in contrast to the *veg* mutant whose internodes decline in concert with those of its normal counterpart as the latter passes through its reproductive phase (22), in the 8-h conditions, internode length in the *gigas* plants remained close to its maximum for at least 130 internodes when the study was terminated. The plants were still strong, vigorous, and of normal appearance at this time. In 8-h conditions, the plants gave the impression that *Sn Dne* activity was proceeding with no effect of age. However, the plants grown in, or transferred to, the 24-h photoperiod gave the impression that some substance

Table II. *Photoperiod response of line 158 (gigas mutant) and its progenitor line 151*

	Node of flower initiation	
Photoperiod (h)	Line 151	Line 158
8 (daylight)	23-24	66 to >130
15 (10 + 5 F,I)*	16-18	43-47
18 (8 + 10 I)*	15	34
24 (8 + 16 I)*	15	Vegetative
Transferred 8 to 24		Transient flowering

*Extension: fluorescent (F) and/or incandescent (I) light.

essential to the flowering process was deficient in these conditions. These observations indicate that the mutant may affect synthesis of this substance rather than sensitivity to it.

In a 24-h photoperiod (8-h daylight + 16-h incandescent light), the F_2 of cross Virtus x mutant segregated clearly into 114 flowering plants and 14 wholly vegetative plants. These numbers are much closer to a digenic 15:1 than a monogenic 3:1. However, the chance of simultaneous mutation at duplicate loci is remote and control by a monogenic recessive with disturbed segregation or incomplete penetration seems likely. F_3 data also point toward single gene control, and the symbol *gi* is provisionally assigned for the *gigas* mutant.

The primary action of this interesting mutant remains unresolved at this stage. The *gigas* mutant appears to be different from all other flowering genes we have encountered in *Pisum*. The mutant does not affect diverse traits like some other mutants, *e.g. sn* and *dm*, and the possibility exists that it is deficient in a fairly specific substance essential to the flowering process, *e.g. gi* may impair synthesis of the floral stimulus.

CONCLUDING COMMENTS

Mutants at these nine major flowering loci are all useful as tools for basic research on the control of flowering. In addition, many of them have value for breeding purposes. Naturally occurring mutants at the *Lf, E, Sn,* and *Hr* loci have already been used extensively to develop cultivars adapted to specific environments and which meet particular requirements (market, canning, freezing, combining, green manure). However, the recent mutants *dne* and *det* have not, as yet, been fully exploited by breeders. Mutant *dne* offers the potential to fine-tune the day-neutral class with possible yield improvements, while *det* could be exploited, in combination with other genes, to reshape the reproductive phase.

Acknowledgments--I thank Dot Steane, Leigh Johnson, and Matthew Ward for technical assistance, Dr. J. B. Reid for helpful comments on the manuscript, and the Australian Research Grants Scheme for financial support.

LITERATURE CITED

1. DŮCHENE C 1984 Reproductive development in *Pisum*. M Sc thesis, Univ of Tasmania
2. GOTTSCHALK W 1982 The flowering behavior of *Pisum* genotypes under phytotron and field conditions. Biol Zbl 101: 249-260
3. GOTTSCHALK W 1988 Phytotron experiments in *Pisum*. 2. Influence of the photoperiod on the flowering behavior of different genotypes. Theor Appl Genet 75: 344-349
4. JOLLY CJ, JB REID, JJ ROSS 1987 Internode length in *Pisum*. Action of gene *lw*. Physiol Plant 69: 489-498
5. MARX GA 1975 The *Le* locus: its influence on flowering time. Pisum Newsl 7: 30-31
6. MARX GA 1986 Linkage relations of *Curl*, *Orc*, and "*Det*" with markers on chromosome 7. Pisum Newsl 18: 45-48
7. MURFET IC 1971 Flowering in *Pisum*: Reciprocal grafts between known genotypes. Aust J Biol Sci 24: 1089-1101
8. MURFET IC 1973 Flowering in *Pisum*: *Hr*, a genes for high response to photoperiod. Heredity 31: 157-164
9. MURFET IC 1975 Flowering in *Pisum*: multiple alleles at the *lf* locus. Heredity 35: 89-98
10. MURFET IC 1982 Flowering in the garden pea: expression of gene *Sn* in the field and use of multiple characters to detect segregation. Crop Sci 22: 923-926
11. MURFET IC 1985 *Pisum* sativum. *In* AH Halevy, ed, Handbook of Flowering, Vol IV. CRC Press, Boca Raton, Florida, pp 97-126
12. MURFET IC 1989 Interaction of the *det* (determinate) mutant with other flowering genes. Pisum Newsl 21 (in press)
13. MURFET IC, S CAYZER 1989 Flowering in *Pisum*: separating genotypes *Sn Hr*, *Sn hr*, and *sn* - on a *lf E Dne* background. Pisum Newsl 21 (in press)
14. MURFET IC, JB REID 1973 Flowering in *Pisum*: evidence that gene *Sn* controls a graft-transmissible inhibitor. Aust J Biol Sci 26: 675-677
15. MURFET IC, JB REID 1974 Flowering in *Pisum*: the influence of photoperiod and vernalizing temperatures on expression of genes *Lf* and *Sn*. Z Pflanzenphysiol 71: 323-331
16. MURFET IC, JB REID 1985 The control of flowering and internode length in *Pisum*. *In* PD Hebblethwaite, MC Heath, TCK Dawkins, eds, The Peak Crop: A Basis for Improvement. Butterworths, London, pp 67-80
17. MURFET IC, JB REID 1987 Flowering in *Pisum*: gibberellins and the flowering genes. J Plant Physiol 127: 23-29
18. REID JB 1979 Red-Far-red reversibility of flower development and apical senescence in *Pisum*. Z Pflanzenphysiol 93: 297-301

19. REID JB 1986 Gibberellin mutants. *In* AD Blonstein, PJ King, eds, Plant Gene Research, Vol. 3: A Genetic Approach to Plant Biochemistry. Springer-Verlag, Wien, pp 1-34
20. REID JB 1989 Gibberellin synthesis and sensitivity mutants in *Pisum*. *In* RP Pharis, SB Rood, eds, Plant Growth Substances 1988. Springer-Verlag, Berlin (in press)
21. REID JB, IC MURFET 1977 Flowering in *Pisum*: the effect of genotype, plant age, photoperiod and number of inductive cycles. J Exp Bot 28: 811-819
22. REID JB, IC MURFET 1984 Flowering in *Pisum*: a fifth locus, *veg*. Ann Bot 53: 369-382
23. SWIECICKI WK 1987 Determinate growth (*det*) in *Pisum*: a new mutant gene on chromosome 7. Pisum Newsl 19: 72-73

MOLECULAR ANALYSIS OF FLORAL INDUCTION IN *PHARBITIS NIL*

Sharman D. O'Neill

Department of Environmental Horticulture, University of California, Davis, CA 95616, USA

One of the most profound events in the life cycle of a plant is the transition from vegetative to reproductive growth. This transition initiates a major re-allocation of assimilates from vegetative biomass production to the production of reproductive organs. For annual plants, this transition also signals the initiation of senescence of the parent plant. In addition to its fundamental importance as a major developmental event in plant growth and development, the reproductive structures of many crop plants, including the major cereal crops, that result from this switch to reproductive growth are of substantial economic importance. This transition from vegetative to reproductive growth has been referred to as floral induction.

Floral induction has been extensively studied for over 50 years with a goal of elucidating the biochemical mechanisms that underlie the process of signaling the transition from vegetative to reproductive growth. This research has relied extensively on photoperiodically-sensitive plants as experimental systems in which floral induction can be rather precisely controlled by the investigator. In general terms, it has been demonstrated that the leaves or cotyledons are the photosensitive tissues and that phytochrome is the photoreceptor of photoperiodically-sensitive plants. Following the appropriate photoperiodic treatment, a translocatable factor moves from the photosensitive tissue to the apex where floral evocation is then triggered (for reviews, see 3, 10).

Many investigators have pondered the nature of the translocatable factor that evokes floral initiation in the apex, yet it has eluded biochemical characterization. Bernier (3) has most recently reviewed the theories that have been advanced to explain the biochemical mechanisms of floral induction. While hypothetical floral hormones (florigen or antiflorigen) have been inferred from grafting experiments, the failure to chemically identify such chemicals suggests that this concept is inadequate. Alternative theories have been advanced that suggest that floral initiation is brought about by alterations in source/sink relations between vegetative tissues and the shoot apex (16), or by multiple levels of control that include plant hormones, nutrients, and perhaps oligosaccharins, polyamines, and secondary plant products being present in the apex at the appropriate concentration and time (4).

Regardless of the nature of the translocatable floral stimulus, it is clear that biochemical processes in the photosensitive tissues of photoperiodically-sensitive plants (*i.e.* leaves or cotyledons) are modulated by photo-treatment in a way that coordinates or modifies the mobilization of factor(s) that are translocated to the shoot apex. The regulation of these biochemical processes may occur at a number of levels, including the level of gene expression. Alterations in patterns of gene expression that may be associated with, or responsible for, floral induction would represent an early step in the transduction of the primary floral stimulus (daylength) to the production of an endogenous factor responsible for floral evocation at the shoot apex (Fig. 1). Our research is directed toward understanding:

1) whether photoinductive treatments result in alterations in patterns of gene expression, and
2) if so, can we clone and test the function of floral-induction-associated genes.

In this paper, I will review the evidence that changes in gene expression may be involved in photoperiodic-dependent production of the floral stimulus and describe the molecular approaches we have initiated to critically assess this hypothesis.

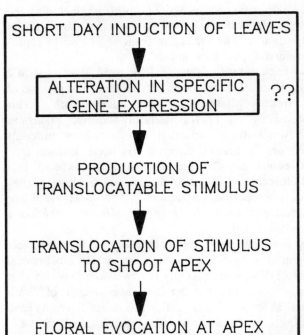

FIG. 1 General model of floral induction in *Pharbitis nil*. Alterations in patterns of gene expression may be associated with, and indeed regulate, production of the signal that evokes flowering at the shoot apex.

PHARBITIS NIL

Pharbitis nil, cv Violet, is a SD plant that has been investigated extensively with regard to floral induction (8, 19). In this sense it represents a well-characterized physiological system. The reasons for its popularity as an experimental organism in which to study SD induction of flowering include:

1) its sensitivity to floral induction at the seedling stage, and
2) its sensitivity to a single cycle of SD photoperiodic treatment.

Pharbitis nil is completely sensitive to SD photoperiodic induction 4 d after germination (14). This corresponds to a seedling stage prior to the appearance of the first true leaves. At this stage, the photo-receptive tissues are the cotyledons, which have been shown by excision experiments to be the source of a translocatable factor responsible for floral evocation at the apex (9). Fifteen days following the photo-inductive treatment, the apex can be scored for its conversion from a vegetative to floral apex. This provides an experimental system in which large numbers of seedlings can be rapidly treated and assessed for floral induction.

A single 12- to 14-h dark treatment is sufficient to completely induce flowering in *Pharbitis nil*, and this induction is blocked by a single 10-min red light exposure 8 h into the dark treatment (18). This extreme sensitivity to photoperiodic treatment provides precise experimental control over the inductive process. Furthermore, the night-break reversal of photo-induction provides an important control to distinguish between nonspecific changes associated with changes in light regime and events truly associated with floral induction.

In addition to the extremely sensitive cultivar of *Pharbitis nil*, cv Violet, we have surveyed a large number of cultivars and related species and found that there is some genotypic variation in the sensitivity to photoperiodic floral induction (Table I). This genetic variability in responsiveness is important in ultimately determining that specific gene expression is functionally related to floral induction. In addition, Epperson and Clegg (6) have recently postulated the presence of an endogenous transposable element in a closely related species, *Ipomoea purpurea*, that affects flower color genes. This suggests that, if similar transposable elements can also be identified in *Pharbitis nil*, we may soon have a molecular tag to isolate specific genetic loci in this species.

Overall, these considerations indicate that *Pharbitis nil* is an excellent experimental system in which to study the molecular mechanisms that underlie photoperiodic floral induction. In the future, we anticipate that genetic transformation of *Pharbitis nil* will be essential to functionally test the role of specific genes in floral induction. Although this has not been accomplished, we are presently establishing the tissue culture requirements for this species that should facilitate its transformation.

Table I. *Sensitivity of Morning Glory Strains to Cycles of SD Induction of Flowering*

Genus	Species	Cultivar	Number of 16-h dark treatments[a]		
			0	1	3
Pharbitis (Ipomoea)	*nil*	Violet	-	+	+
		Kidachi	-	+	+
		Tendan	-	+	+
		Scarlet O'Hara	-	-	N.D.
		Showy Climber	-	+	N.D.
		Alf 8	-	-	N.D.
		C.A. #63.8	-	-	+/-
		C.A. #62.76	-	+	+
Ipomoea	sp.	Early Call	-	+	N.D.
	tricolor	Heavenly Blue	-	+	N.D.
	hederacea	Roman Candy	-	+	+
	purpurea	U.S. species	-	-	-
	purpurea	Argentina species	-	-	-
	hederacea	U.S. species	-	-	-
	hederacea	Iran species	-	+	+

[a] The indicated number of 16-h dark cycles were given, and transition of the apex from vegetative (-) to floral (+) growth was assessed. N.D. = not determined.

THE ROLE OF GENE EXPRESSION IN FLORAL INDUCTION

Inhibitor Studies. The role of changes in patterns of gene expression in relation to floral induction was first addressed by the examination of the effects of inhibitors of transcription and translation on the induction process (2, 5, 7, 15, 17, 20, 21). The results of several early experiments were contradictory, with some indicating that changes in either transcription or translation were not required for floral induction and others indicating that these processes were required for induction (Table II).

In general, experiments with inhibitors were carried out by application of the inhibitor to the photosensitive tissue and, following the photo-inductive treatment, examining the effect of the inhibitor on floral evocation. In addition to the normal caveats associated with the interpretation of inhibitor effects, namely that inhibitors may fail to penetrate tissues and that they certainly have nonspecific effects, these experiments were further complicated in that the inhibitor may have been translocated from the photosensitive tissue to the shoot apex. Thus, it is difficult to determine whether any inhibitor effect was related to production of a translocatable factor or to its perception in the apex.

Table II. *Compilation of the Effects of Inhibitors of Transcription and Translation on Floral Induction in Photoperiodically-Sensitive Plants*

	Inhibition of floral induction
Perilla, SDP (Zeevart, 1969)	
5-Fluorouracil	NO
2-Thiouracil	NO
Ethionine	NO
Lolium, LDP (Evans, 1969)	
p-Fluorophenylalanine	NO
Chloramphenicol	NO
Ethionine	NO
5-Fluorouracil	SLIGHT
Xanthium, SDP (Collins et al., 1963)	
p-Fluorophenylalanine	YES
2,6-Diaminopurine	YES
Benzimidazole	YES
Picolinic acid	YES
Quercitin	YES
Methyl-methionine	YES
2-Thiouracil	YES
Xanthium, SDP (Ross, 1970)	
Cycloheximide	YES
Pharbitis nil, SDP (Arzee et al., 1970)	
Actinomycin D	YES

Protein Changes. A more direct examination of the role of gene expression in floral induction has involved the examination of protein profiles in induced and noninduced cotyledons of *Pharbitis nil*. An early study separated proteins by isoelectric focusing and concluded that the abundance of one polypeptide decreased under photoinductive conditions (SD) (17). Later studies examined profiles of proteins extracted from *Pharbitis nil* cotyledons subjected to LD, SD, or night-break treatments by two-dimensional PAGE (12). In this study, several changes in polypeptide abundance associated with floral-inductive treatment were observed, with the most pronounced change being a decrease in the abundance of four polypeptides.

RNA Changes. Yoshida et al. (20) examined RNA content of induced and noninduced cotyledons of *Pharbitis nil*. In this study, RNA was separated into various classes by chromatography on methylated serum albumin and the base content of each fraction determined. At this level of analysis, it was observed that the G+C content of the mRNA fraction decreased somewhat upon induction, leading the investigators to conclude that changes in mRNA composition accompanied photoperiodic induction in the cotyledon of *Pharbitis nil*. Extrapolating from this conclusion, Yoshida et al. (20) proposed that SD treatment resulted in derepression of a specific gene which gave rise to a specific floral stimulus mRNA (termed FS-RNA) that encoded a protein required for synthesis of the translocatable factor involved in floral induction.

Recent studies have examined the appearance of specific mRNA species in far greater detail (12). In this work, poly(A)+RNA was isolated from induced and noninduced cotyledons of *Pharbitis nil* and used to program *in vitro* translations and the polypeptides analyzed by PAGE. At the level of one-dimensional gels, no changes in mRNA composition were apparent between SD-induced cotyledons and noninduced cotyledons subjected to a night-break treatment. However, when analyzed on two-dimensional gels, a single mRNA encoding a 28-kD polypeptide was observed to be quantitatively increased in SD relative to night-break treated cotyledons (Fig. 2; 12). These results have provided the most direct evidence that changes in patterns of gene expression are associated with, and may indeed be responsible for, photoperiodic induction of flowering in *Pharbitis nil*. However, these results also indicate that the changes in gene expression are extremely subtle, with the most apparent changes involving quantitative regulation of a relatively low-abundance transcript.

CLONING OF FLORAL STIMULUS mRNAS

In order to define the function of genes whose expression changes upon photoperiodic floral induction, it is necessary to clone these genes, examine their regulation in detail, and ultimately subject them to a functional test in transgenic plants. To begin this analysis, we have attempted to clone cDNAs whose pattern of expression is altered upon SD photoperiodic treatment of *Pharbitis nil* cotyledons by differential screening of cDNA libraries.

Because results of *in vitro* translation of isolated mRNA indicated that changes in mRNA composition may be quite subtle (Fig. 2; 12), DNA libraries were constructed in a vector that would facilitate subsequent enrichment strategies for the identification of rare transcripts. Several cDNA libraries were constructed by a vector-primer method of Alexander (1), using a modified vector pCGN1703 (Fig. 3). The cDNA cloning in this vector is directional, with a bacteriophage T7 RNA polymerase promoter adjacent to the 5' end of the cDNA. The plasmid vector also has the intergenic region of M13 and, thus, can be isolated as single-stranded DNA in phage particles. Together, these features allow unique opportunities for enrichment of mRNA-derived cDNA probes and for enrichment of the cDNA libraries.

We have initiated screening of a SD-induced cotyledon cDNA library by differential hybridization to [^{32}P]-labeled cDNA probe derived from either

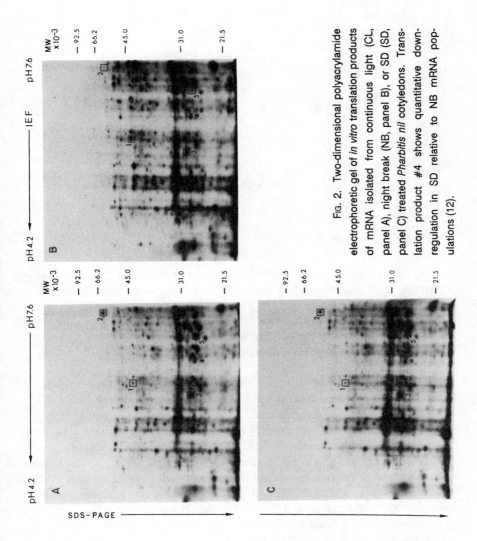

FIG. 2. Two-dimensional polyacrylamide electrophoretic gel of in vitro translation products of mRNA isolated from continuous light (CL, panel A), night break (NB, panel B), or SD (SD, panel C) treated *Pharbitis nil* cotyledons. Translation product #4 shows quantitative down-regulation in SD relative to NB mRNA populations (12).

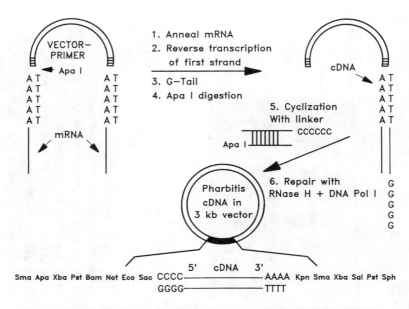

FIG. 3. Schematic diagram of the construction of *Pharbitis nil* cDNA libraries in pCGN1703. Following annealing of the mRNA to homopolymer T-tailed vector, a first-strand cDNA is synthesized, G-tailed, the plasmid recircularized with synthetic linker, and the second strand synthesized *in vitro* with RNaseH and DNA polymerase I. The library plasmid has a T7 RNA polymerase promoter at the 5' end of the cDNA and contains the intergenic region of M13 for production of single-stranded library DNA.

continuous light or SD-induced cotyledon mRNA. Several rounds of screening have indicated that, at this level of sensitivity, very few differentially expressed cDNAs can be detected. Typical screens of greater than 10^5 cDNA clones have failed to reveal any qualitatively up-regulated cDNAs. We have identified one cDNA (pSD13) whose expression appears to be quantitatively down-regulated under SD treatment (Fig. 4) and have tentatively identified two quantitatively up-regulated cDNAs. These results clearly indicate that the changes in gene expression that accompany floral induction are not dramatic. This can be clearly contrasted with changes in gene expression that accompany flower senescence (11) or fruit ripening (13) where the expression of many abundant (>0.5%) transcripts is activated.

We anticipate that the isolation of qualitatively regulated genes involved in floral induction will require, at a minimum, significant probe- and library-enrichment strategies. However, truly regulatory gene products may be present at such low levels that genetic approaches may be required to clone these genes. To this end, we are developing long-term strategies employing transposable elements to tag genes involved in photoperiodic floral induction in *Pharbitis nil*.

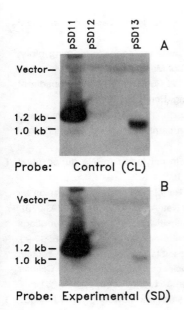

FIG. 4. Reverse Northern of cDNA clones isolated by differential hybridization. cDNAs were isolated from purified bacterial colonies, size fractionated on an agarose gel, blotted to nitrocellulose, and probed sequentially with [^{32}P]-labeled cDNAs derived from continuous light (panel A) or SD (panel B) mRNA. Lane 1: constitutive cDNA (pSD11). Lane 4: quantitatively downregulated cDNA (pSD13).

Acknowledgments--Many thanks to Drs. R.M. Sachs and M.S. Reid for helpful discussions, to R.A. Bicknell for laboratory assistance, and to Dr. D. Alexander (Calgene, Davis, CA) for providing the plasmid pCGN1703. This research was supported by USDA/CRGO, Grant No. 87-CRCR-1-2420.

LITERATURE CITED

1. ALEXANDER DC 1987 An efficient vector-primer cDNA cloning system. *In* R Wu, L Grossman, eds, Methods in Enzymology, Vol. 154, Recombinant DNA, Part E. Academic Press, New York, pp 41-64
2. ARZEE T, J GRESSEL, E GALUN 1970 Flowering in *Pharbitis*: the influence of actinomycin D on growth, incorporation of nucleic acid precursors, and autoradiographic patterns. *In* G Bernier, ed, Cellular and Molecular Aspects of Floral Induction. Longman Ltd, London, pp 93-107
3. BERNIER G 1988 The control of floral evocation and morphogenesis. Annu Rev Plant Physiol Plant Mol Biol 39: 175-219
4. BERNIER G, J-M KINET, RM SACHS 1981 The Physiology of Flowering, Vol. I, The Initiation of Flowers, Vol. II, The Transition to Reproductive Growth. CRC Press, Boca Raton, FL
5. COLLINS WT, FB SALISBURY, CW ROSS 1963 Growth regulators and flowering. III. Antimetabolites. Planta 60: 131-144
6. EPPERSON BK, MT CLEGG 1987 Instability at a flower color locus in morning glory. J Heredity 78: 346-352
7. EVANS LT 1969 *Lolium temulentum* L. *In* LT Evans, ed, The Induction of Flowering. Some Case Histories. MacMillan, Melbourne, pp 328-349

8. IMAMURA SI (ed) 1967 Physiology of Flowering in *Pharbitis nil*. Japanese Soc Plant Physiol, Tokyo, Japan
9. IMAMURA S, A TAKIMOTO 1955 Photoperiodic responses in Japanese morning glory, *Pharbitis nil* Chois., a sensitive short day plant. Bot Mag Tokyo 68: 235-241
10. LANG A 1965 Physiology of flower initiation. *In* W Ruhland, ed, Encyclopedia of Plant Physiology, Vol. 15 (Part 1). Springer-Verlag, Berlin, pp 1380-1536
11. LAWTON KA, B HUANG, PB GOLDSBROUGH, WR WOODSON 1988 Molecular cloning and characterization of petal senescence genes from carnation flowers. Plant Physiol 86S: 16
12. LAY-YEE M, RM SACHS, MS REID 1987 Changes in cotyledon mRNA during floral induction of *Pharbitis nil* strain Violet. Planta 171: 104-109
13. LINCOLN JE, S CORDES, E READ, RL RISCHER 1987 Regulation of gene expression by ethylene during tomato fruit development. Proc Natl Acad Sci USA 84: 2793-2797
14. MARUSHIGE K, Y MARUSHIGE 1963 Photoperiodic sensitivity of *Pharbitis nil* seedlings of different ages in special reference to growth patterns. Bot Mag Tokyo 76: 92-99
15. ROSS C 1970 Antimetabolite studies and the possible importance of leaf protein synthesis during induction of flowering in the cocklebur. *In* G Bernier, ed, Cellular and Molecular Aspects of Floral Induction. Longman Ltd, London, pp 139-151
16. SACHS RM, WP HACKETT 1983 Source-sink relationships and flowering. *In* WJ Meudt, ed, Strategies of Plant Reproduction, BARC Symp No. 6, Allenheld, Osmun, Totowa, NJ, pp 263-272
17. STILES JI, PJ DAVIES 1976 Qualitative analysis by isoelectric focusing of the protein content of *Pharbitis nil* apices and cotyledons during floral induction. Plant Cell Physiol 17: 855-857
18. TAKIMOTO A, KC HAMNER 1965 Studies on red light interruption in relation to timing mechanisms involved in the photoperiodic response of *Pharbitis nil*. Plant Physiol 40: 852-854
19. VINCE-PRUE D, J GRESSEL 1986 *Pharbitis nil*. *In* AH Halevy, ed, Handbook of Flowering, Vol. IV. CRC Press, Boca Raton, FL, pp 47-81
20. YOSHIDA K, K UMEMURA, K YOSHINAGE, Y OOTA 1967 Specific RNA from photoperiodically induced cotyledons of *Pharbitis nil*. Plant Cell Physiol 8: 97-108
21. ZEEVAART JAD 1969 *Perilla*. *In* LT Evans, ed, The Induction of Flowering. Some Case Histories. MacMillan, Melbourne, pp 116-155

GIBBERELLINS AND FLOWERING IN HIGHER PLANTS: DIFFERING STRUCTURES YIELD HIGHLY SPECIFIC EFFECTS

RICHARD P. PHARIS, LLOYD T. EVANS, RODERICK W. KING, AND LEWIS N. MANDER

Biological Sciences Department, University of Calgary, Calgary, Alberta T2N 1N4, Canada (R.P.P.); *Division of Plant Industry, C.S.I.R.O.* (L.T.E., R.K.), *and Research School of Chemistry, Australian National University* (L.N.M.), *Canberra, Australia*

This review covers recent work dealing with gibberellic acids (GAs) and flowering in selected species of gymnosperms, woody angiosperms, and SD and LD herbaceous angiosperms. Other species and phytohormones are not covered, and for these, the reader is referred to (1, 6-8, 13, 22, 24 and references cited in these reviews).

PROMOTION OF FLOWERING IN CONIFERS BY GIBBERELLINS

The possible causal roles of GAs in promoting precocious and enhanced flowering in conifers have been discussed in detail in several reviews (4, 14, 16, 17, 19). Application of several bioactive GAs, including GA_3, promotes both precocious and normal flowering in the Cupressaceae and Taxodiaceae, and this flowering can be synergistically enhanced by the use of cultural treatments such as water stress, rootpruning, cambial girdling, etc. (17). However, for conifers of the Pinaceae, only the less-polar GAs (*e.g.* $GA_{4/7}$ and $GA_{4/7}$ + GA_9) repeatably promote flowering in both juvenile and older propagules. As for the other families, these GAs are most effective when given with adjunct cultural treatments. In fact, for the Pinaceae, over 20 species in five genera flower in response to $GA_{4/7}$.

The flowering response to exogenous application of GAs may imply that endogenous GAs play a part in the flowering process, but does not prove it. Support for a causal role for endogenous less-polar GAs in conebud differentiation is provided by Table I where rootpruning was used to promote flowering of Douglas fir (*Pseudotsuga menziesii*) over 20-fold in the absence of exogenous hormone application. Here, less-polar GAs (GA_4, GA_7, and GA_9) and the polar GA (GA_3) have been quantified by bioassay. All are native GAs in Douglas fir. Extracts of the shoots (minus needles) on which the conebud primordia are forming showed that the rootpruning treatment had increased the concentration and/or amount of less-polar GAs by 2- to 4-fold, while leaving the more-polar GA_3 either unaffected (on a concentration basis), or diminished (on a per shoot basis). Similar results from other cultural treatments are noted for several other conifer species in Pharis and Ross (16), and Pharis et al. (17).

Metabolism of the native GA_4 was also followed (Table I) over 192 h during that period when biochemical differentiation of strobili (male and female conebuds) occurs. The highly effective rootpruning treatment retarded [^{3}H]GA_4 metabolism (Table I). A similar phenomenon was noted for the native GA_9, also fed in [^{3}H] form (3) to Picea abies propagules that were subjected to a high temperature cultural treatment.

Table I. *Effect of rootpruning on flowering and on the endogenous GA-like substances, and proportion of unmetabolized (^{3}H)GA_4 in elongating shoots from 5- to 10-year-old half-sibling Douglas-fir trees near Victoria, B.C., Canada*[a]

Treatment[b]	[^{3}H]GA_4[e] (% total radioactivity in free GA fraction)	Female flowering	
		on adjacent branches of trees analyzed for GAs	in overall experiment
Control	28%	0	24
Rootpruned	45%	4+[f]	507

Treatment[b]	Less-polar endogenous GAs[d]				Polar GAs[d]	
	$GA_{4/7}$-like[e], ng GA_3 equiv. per:		$GA_{9/15}$-like[e], ng GA_3 equiv. per:		GA_3[e] ng per:	
	shoot	g dry wt	shoot	g dry wt	shoot	g dry wt
Control	2.3	2.8	13.8	10.3	15.0	11.2
Rootpruned	9.2	13.2	13.0	18.5	7.8	11.0

[a]Adapted from Pharis et al. (17) and Ross et al. (20).

[b]Saplings were rootpruned in early April; tissue was harvested in mid-June for analysis of GA-like substances by bioassay and gas chromatography-selected ion monitoring (GC-SIM).

[c] [^{3}H]GA_4 (Sp. Act., 1.4 Ci/mmol); was approximately 100 nCi/shoot.

[d]GA-like substances bioassayed on dwarf rice, cv Tanginbozu, after C18 reversed-phase HPLC in serial dilution.

[e]GA_3, GA_4, GA_7 and GA_9 were identified by capillary GC-SIM (B. Thompson, et al. unpublished).

[f]Rated on a scale of 0 (absent) to 5 (abundant).

How might these cultural treatments influence endogenous GA levels? A common feature may be plant water status. First, trees from which data in Table I are derived were assessed for water status (23), and indeed root-pruning significantly increased water stress in the shoot tissue, relative to control, unpruned trees [interestingly, exogenous application of $GA_{4/7}$ alone to other trees in the factorial study reduced water stress, relative to control shoots, and gave enhanced flowering, but only on fecund clones (23)]. A similar effect on plant water status might be expected for high temperature, an effective floral-promotive treatment in spruce and hemlock (17). It is thus possible that water stress (and at least certain of the other cultural treatments) interferes with the metabolism (hydroxylation and/or conjugation) of native GAs, resulting in a buildup of specific GAs, most notably those without hydroxyls, or with only one hydroxyl group. In fact, a similar trend toward a greater concentration of less-polar GAs was noted for a *Cupressus* species subjected to N starvation (which promotes juvenile flowering), even though that genus will flower in response to application of the polar GA_3 (cited in 17).

It has thus been postulated (16, 17) that conifers regulate their flowering in nature through this mechanism. Indeed, early flowering of juvenile trees and more abundant flowering of mature trees occurs naturally in years following hot, dry conditions during the previous spring/summer when conebud initiation and early differentiation are occurring. Additionally, evidence from application of $GA_{4/7}$ to Douglas-fir genotypes of varying fecundity (23) showed that families with a history of 'poor flowering' produced significant elongation, but nil or minimal flowering, whereas families with a history of 'good flowering' showed no significant increase in shoot elongation, but did flower significantly. Speculatively, a 'poor flowering' family may be one which is GA deficient--it may use the $GA_{4/7}$ first for vegetative growth, whereas a family with a history of 'good flowering' may have endogenous GA concentrations which are already optimal for vegetative shoot elongation. In juvenile *Cupressus* seedlings of differing age, low dosages of GA_3 yielded only increased vegetative growth in seedlings of all ages but, as dosage was increased, older seedlings flowered first/best (cited in 14). Thus, older trees, trees of fecund genotypes, and trees subjected to environmental stress all may have elevated levels of endogenous less-polar GAs which are highly 'florigenic' in the conifers.

DO GAs POSSIBLY ACT BY DIVERTING NUTRIENTS TO THE POTENTIAL CONEBUD?

Sachs (cited in 1) postulated that flowering may be brought about simply by the diversion of nutrients, and that florigenic factors act primarily by ensuring that the shoot apex receives a greater supply of assimilates under induced, than under non-induced conditions.

This postulate was tested with the *Pinus* system using Radiata pine (*Pinus radiata*) and $GA_{4/7}$ treatment. $^{14}CO_2$ was fed to needles subtending the shoot on day 8 after the hormone treatment (Table II). Here, treatment of the shoot of mature propagules with $GA_{4/7}$ during the conebud differentiation period significantly enhanced the movement of [^{14}C]assimilates to lateral long-shoot primordia (next year's potential female conebuds or lateral vegetative shoots). A similar effect of $GA_{4/7}$ was noted for dry matter allocation within the shoot (18 and Table III).

However, when GA_3, which is relatively ineffective for flowering in Pinaceae, was tested in comparison with the floral-promotive $GA_{4/7}$, both GAs significantly mobilized assimilates to lateral primordia undergoing differentiation, yet only the less-polar $GA_{4/7}$ significantly promoted flowering (Table III). This implies a morphogenic effect for $GA_{4/7}$ over and above that of assimilate diversion.

Table II. *Assimilate partitioning in Pinus radiata shoots: effects of $GA_{4/7}$ on ^{14}C photoassimilate allocation*[a]

	Treatment		Difference	
Bud part	Control (nCi)	$GA_{4/7}$ (nCi)	%	Signif.
Entire bud	64.22	60.97	-5.1	n.s.
Apical dome	0.32	0.29	-10.2	n.s.
Long-Shoot Primordia[b]	5.90	8.49	43.9	0.05
Structural tissues	58.00	52.18	-10.0	n.s.

[a]Adapted from Ross *et al.* (18). $^{14}CO_2$ applied to needles 7 d after shoots treated with $GA_{4/7}$. Shoots analyzed for ^{14}C content 24 to 40 h after $^{14}CO_2$ application.
[b]Long-shoot primordia are capable of differentiating into either vegetative lateral buds or female conebuds (strobili).

GIBBERELLINS AND FLOWERING IN WOODY ANGIOSPERMS

Unlike the Coniferae, GAs have generally had no effect or have even inhibited flowering in many species of woody angiosperms (see 1, 13, 25). In certain species/varieties of fruit trees, including apple, pear, apricot, and citrus, there is a severe bienniality in flowering/fruiting, and the evidence is rather compelling that the reason for this is the movement of bioactive GAs from the developing fruit into the adjacent spur apex during floral initiation/early differentiation (10, 13). A similar situation may exist for *Ribes*, except that the source of the bioactive GAs appears to be the roots (21). There are, however, some instances of applied GA_3 initiating floral development in woody angiosperms (cited in 13), but these have not been repeatably

Table III. *Dry matter partitioning[1] in Pinus radiata shoots: a comparison of the effects of GA_3 vs $GA_{4/7}$[2]*

Treatment	Cone[4] buds/shoot (no.)	For all long-shoot primordia[3] dry wt (mg)	Average of each long-shoot primordium dry wt (μg)
Control	1.0 a[5]	1.09 a	97 a
GA_3	1.7 a	2.59 b	231 b
$GA_{4/7}$	4.0 b	2.50 b	215 b

[1] Samples for dry weight determinations were taken at 2 to 4 d intervals from Febryary 9 to March 16, and weights shown are averages for all sampling dates. $GA_{4/7}$ and GA_9 treatments were applied on February 8, 13, and 22.
[2] Adapted from Ross et al. (18).
[3] Long-shoot primordia are capable of differentiating into either vegetative lateral buds or female conebuds (strobili).
[4] Assessed the next spring for propagules from which long shoot buds had been sampled the previous year.
[5] Values followed by the same letter do not differ significantly at $p < 0.05$.

documented. There are also several instances (cited in 10) where GA_3 will not inhibit flowering in a species (apple) which is well known for its floral-inhibitory response to application of GA_3, $GA_{4/7}$, and GA_7.

Recently, application of GA_4 (which is native to apple, and is a rapidly metabolizable monohydroxylated GA lacking a double bond in ring A) has been shown to promote return bloom in spur varieties of apple, and an example of this is shown in Table IV. Under similar circumstances, applications of GA_3 and GA_7 are inhibitory to flowering (cited in 10, 13, 15). The positive effects on return bloom have also been obtained with C-3 epi-GA_4 (10), and with the C-13 acetate form of GA_1 (15). However, GA_4 has not promoted precocious flowering in any woody angiosperm, nor did it promote flowering under non-inductive conditions in *Fuchsia* (R. Sachs, personal communication).

Why are the GAs which are highly bioactive for vegetative growth so inhibitory to flowering in woody angiosperms, whereas GA_4 can actually promote flowering (*e.g.* return bloom)? One explanation is that different processes are being regulated. Goldschmidt and Monselise (5) suggested that flowering in citrus was regulated (negatively) by GAs, the assumption being that floral induction had occurred, but that subsequent events depended upon the concentration of GAs being below a certain threshold. Similarly, for apple, Pharis and King (13), drawing on an analogy by Buban and Faust (2), speculated that applied highly bioactive GAs may, as do other floral inhibitory treatments (high temperature, heavy cropping of this year's fruit), reduce the number of nodes on the spur, thereby lengthening the plastochron and suppressing flower bud formation on the older organ primordia. The promotive effect by GA_4 (Table IV) could be due to enhancement of early floral differentiation in induced apices of at least some spurs, and rapid metabolism of the GA_4 may prevent it from acting in the inhibitory manner so notable for the more slowly metabolized GA_3 (and probably also GA_7).

SHORT DAY PLANTS -- WHAT ROLE, IF ANY, DO GAs PLAY IN FLORAL INDUCTION/INITIATION?

As in woody angiosperms, the application of bioactive GAs to a wide range of SD plants will generally inhibit flowering under inductive conditions, and does not promote it under noninductive conditions (see 1, 13, 24 and references cited therein). However, there are a few SD plants where GAs will promote flowering under noninductive and marginally inductive conditions (1, 9, 13, 24). An example of such promotion is given in Figure 1 for plants of a dwarf cultivar of *Pharbitis nil* treated with three different GAs at two different times before a marginally-inductive long night.

Here, GA_3 (especially) can be shown to exert the usual negative (inhibitory) effect on flowering, but only when applied at rather high doses. At lower doses even GA_3 can be promotive, as can GA_4 over a rather wide range of doses (Fig. 1). GA_1, the proposed 'effector' for vegetative growth (see 11, and ref. cited therein), becomes promotive of flowering only at very high doses. GA doses inhibitory to flowering also caused a significant elongation of the stem (9), although the elongation brought about by GA_1 (9) was not accompanied by inhibition of flowering (the flowering response to GA_1 is shown in Fig. 1b).

Gibberellins, then, can play a positive role in floral induction of SDP, even the qualitative SD plant *P. nil* (Fig. 1). However, that role is only seen

Table IV. *The effect (as percent of flowering spurs) of GA_4 and of GA_4 + zeatin, applied to defruited spurs on nine Golden Delicious apple trees, on the flowering of these spurs the next year.*

Each tree was in its "on" year and was hand-thinned to one fruit on every fifth flowering spur. Each of the nine treatments was applied to 25 spurs per tree.

Treatment (Mg/spur)	Treatment time (weeks after full bloom)	
	4.5	7
Control (0; 60% ethanol)	2.27 a[2]	
GA_3 (3)	7.74 b	2.37 a
GA_4 (30)	6.21 b	3.95 ab
GA_4 (30) + zeatin (30)	8.22 b	6.82 b
GA_4 (300)	12.67 c	13.71 c

[1]Adapted from Looney *et al.* (10) with permission from Springer-Verlag, Berlin.
[2]Mean values followed by differing letters differ significantly at $p < 0.05$ using Duncan's Multiple Range Test: SE - 1.37.

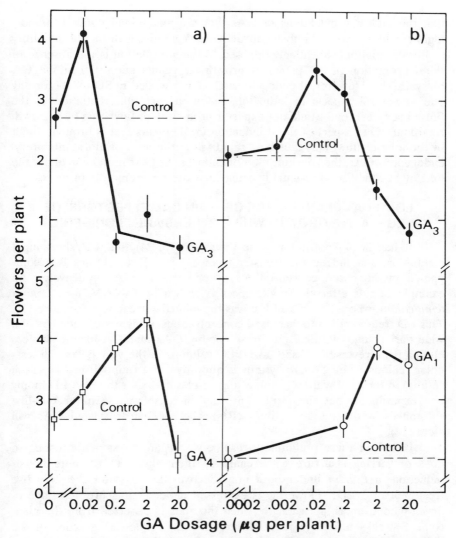

FIG. 1. Effect on flowering of varying amounts of GAs applied once to the petiole of *Pharbitis nil*, cv 'Kidachi' either 5 h (a) or 16 h (b) before starting an inductive 13.25 h dark period. Ethanol concentration was 95% for (a) and 80% for (b).

when using a somewhat GA-deficient dwarf cv (see 9), and under marginally inductive long nights. Even under these limiting circumstances, GAs can be highly inhibitory, especially at high doses (Fig. 1), or when applied immediately after the inductive long night, or when the GA applied is long-lived, such as GA_3 (9).

Work on endogenous GAs in this SD response system has yet to be published. However, as for woody angiosperms, it is possible to speculate that the high concentrations of GAs that are maintained under LD/short night conditions are not only highly active in vegetative growth, but inhibitory to floral initiation. Seemingly opposed to this postulation is the absence of floral promotion in SD plants by growth retardants known to inhibit GA biosynthesis. However, if endogenous GAs are needed in SD plants for any one or several aspects of induction, initiation/early differentiation of the floral apex, then one would not expect a promotion of flowering by a growth retardant. One aspect of floral induction in SD plants that is brought about by the long night(s) may thus be a rapid fall in the level of 'floral inhibitory' endogenous GAs, the concentration of endogenous GAs remaining under the long night conditions presumably being adequate for promotory purposes.

LONG DAY PLANTS -- ARE CERTAIN GAs 'EFFECTORS' OF VEGETATIVE GROWTH WHILE OTHERS ARE FLORIGENIC?

It has been known for some time (see 13, 24) that LD/short night conditions, and indeed the transfer from SD to LD conditions is accompanied by increased concentrations of endogenous GAs in leaves. An example of this effect in shoot apices is shown in Table V, for *Lolium temulentum*, where a single 24-h d was the inductive treatment (see also 12). This LD treatment led to increased concentrations of bioactive-free GAs of a wide variety of polarities, but most notably GAs and GA-like substances eluting from reversed phase C18 HPLC columns in the very polar and less-polar regions. The *Lolium* system is quite dynamic (unpublished research results) in terms of which GAs may appear elevated after the 24-h LD in any one experiment, but the most consistent increase was seen for GA-like substances at the retention times (Rts) where polyhydroxylated GAs will elute (12).

Based on these preliminary results, we began a systematic testing of GAs of varying structure with regard to their ability to promote flower induction in *Lolium* under both noninductive SD and after the inductive treatment of 1 LD. Flowering is expressed either as a 'score' for stage of development, or as 'apex length' when the apex is dissected from the plant some 3 weeks after the inductive treatment. A flowering score of 2.0 indicates the presence of double ridges, the earliest unambiguous sign of inflorescence initiation.

In *L. temulentum* leaves, both GA_1 and GA_3 are native, as are a wide variety of other GAs (D Pearce *et al.*, unpublished), but when these polar, highly bioactive (in terms of vegetative elongation growth) GAs are applied to *L. temulentum* leaves under noninductive SD, doses of about 5 µg of GA_3 and >50 µg of GA_1 are required to reach a flowering score of about 2 (Fig. 2).

However, if one 'builds' a more hydroxylated GA molecule by, for example, adding a β-hydroxyl group to GA_3 at C-15, florigenic activity of the

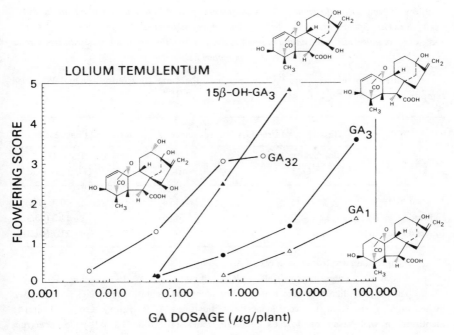

FIG. 2. Effect of dose for several GAs on the flowering response of *Lolium temulentum* plants. Flowering was assessed 3 weeks after inductive treatment when apices were dissected out (2 = double ridges present, 4 = glume primordia present, 5 = lemma present). Plants were held in noninductive SD at 25/20°C. GA_1 (△), GA_3 (●), 15β-OH-GA_3 (▲), GA_{32} (○). Adapted from Pharis *et al.* (12) and LT Evans, RW King, LN Mander, and RP Pharis, (unpublished).

molecule is enhanced 10- to 20-fold (*e.g.* 5 μg GA_3 = 0.3 μg 15β-OH GA_3, Fig. 2). Similarly, the minimum effective dose of the polyhydroxylated GA_{32} is approximately 1/1000 of that of GA_1 (Fig. 2). Thus, the appearance in the *Lolium* apex of endogenous GA-like substances of a putative polyhydroxylated nature following 1 LD (Table V) seems likely to be associated with floral evocation.

An even more telling index of the florigenic activity of applied GAs can be obtained by measuring not only the flowering score (or apex length), but also length of the stem. In Figure 3, it can be seen that response to 1 LD gives a large promotion in apex length with little increase in stem length at the time of dissection. Conversely, GA_1 (at 50 μg per plant) gives a 3- to 4-fold increase in stem length, with little increase in apex length (Fig. 3). However, GA_{32}, the polyhydroxylated GA with a C-1,2 double bond, gives a flowering response at 0.5 μg per plant which approaches that obtained from an 18-h LD, while yielding only a nominal increase in stem length (Fig. 3).

Table V. *Changes in endogenous GA-like substances of a polyhydroxylated (at or near the Rt of GA_{32} and GA_8), polar (at or near the Rt of GA_1) or less polar (at or near the Rts of GA_{20}, GA_4, GA_9, GA_{12} or kaurenoic acid) nature in shoot apices of* L. temulentum *associated with exposure to a single inductive LD.*[a]

Figures in parentheses are pg per apex, all others are $\mu g\ g^{-1}$ dry weight. 1983-1984 samples, experiments 230, 231.

	Bioactivity of Free GA-like Substances at C_{18} Reversed Phase HPLC		
	Polyhydroxylated GAs	Rt of Polar GAs[b]	Less Polar GAs[c] (incl. GA precursors)
Control (SD)	2.6 (9.5)	12.6 (46.4)	0.3 (1.2)
Induced (Day II)	12.5 (47.6)	11.4 (43.5)	11.3 (43.3)
Induced (Day III)	3.0 (11.3)	1.2 (4.6)	11.7 (44.2)

[a]Adapted, in part, from Pharis *et al.* (12)

[b]GA_1 was not detectable by capillary GC-SIM.

[c]The only GA-like substance that could be identified (tentatively) by GC-SIM from the bioactive fractions from induced apices was putative 11 μ-hydroxy GA_7. Tentative identification was based on Kovat's retention index (KRI) and a full mass spectrum (unpublished data, D Pearce, R Pharis, LN Mander, L Evans, and R King).

This differential effect on flowering (apex length) versus vegetative elongation (stem length) was also apparent in a further experiment shown in Figure 4. GA_1 (with hydroxyls at C-3 and C-13, but with no double bond in ring A), gives poor flowering but good vegetative growth. Simply adding a double bond (C-1,2) in ring A, as in GA_3, increases the 'florigenic' effect, but also promotes stem elongation. Adding a third hydroxyl group at C-15 (15β-OH-GA_3) gives a greatly increased 'florigenic' effect at lower doses, at which there is only a small effect on stem length.

In essence, then, 'building' towards the GA_{32} structure also builds toward the effect obtained with a single inductive LD!

Results from the LD *Lolium temulentum* system are still preliminary. Nonetheless, they indicate, as noted in (12), a specificity of function with regard to GA structure in floral induction/initation that differs from that in the promotion of shoot elongation.

CONCLUSIONS

The examples given most recently for the *Lolium temulentum* LD-induced flowering system best show the differing specificity of function with regard to GA structure. Even so, a not dissimilar analogy exists for the SD-induced *Pharbitis nil* system (albeit tested under marginally inductive long

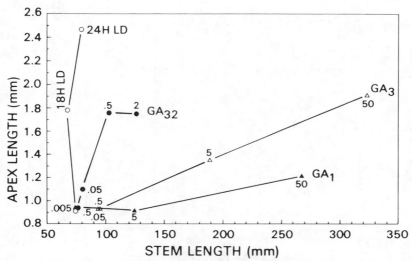

FIG. 3. Relative effect of several GAs on flowering response (shoot apex length) and growth response (stem length) of *Lolium temulentum* plants held in SD (O), compared with those of plants exposed to one LD of either 18 or 24 h duration. The curves join points for each step in the dose responses (Fig. 2) for each gibberellin. GA_1 (▲), GA_3 (△), GA_{32} (●) were applied to a single leaf at the doses (per plant) indicated. Adapted from Pharis *et al.* (12).

nights), with GA dosage and timing of application governing promotion versus inhibition (9). Additionally, the promotive effects of GA_4 (C-1,2 dihydro) in promoting return bloom in apple, contrasted with only inhibitory effects obtained by either GA_3, or especially GA_7 (both C-1,2 dehydro), imply differential effects (again, promotion versus inhibition) for differing GA structures on the initiation/early differentiation phases/processes of flowering in woody angiosperms. Finally, the differential effects between GA_3 and the $GA_{4/7}$ mixture in promoting flowering in Pinaceae species are correlated well with endogenous GA concentrations and GA metabolism in plants where flowering in these woody perennials is promoted by cultural (stress) treatments. The very similar efficacies in assimilate partitioning within the *Pinus* shoot for GA_3 and $GA_{4/7}$, but dissimilarity for flowering (only $GA_{4/7}$ significantly promoted flowering), implies again that floral morphogenetic effects of certain GAs may be separable from their other actions.

Specific effects from GAs of differing structure thus seem possible, indeed likely. However, subsequent oxidative metabolism and/or differential sensitivity of a specific organ could still provide a mechanism by which the organ (*e.g.* the apex) in many plants could 'recognize' one or several GAs of a common structure for the initiation of rather diverse processes.

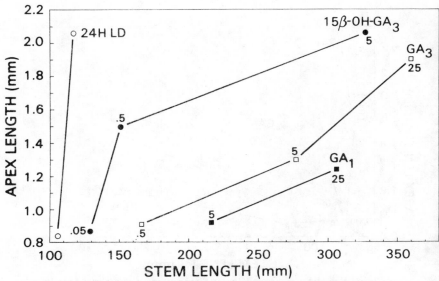

FIG. 4. Relative effect of several GAs on flowering response (shoot apex length) and growth response (stem length) of *Lolium temulentum* plants held in SD (O), compared with those of plants exposed to one LD of 24 h (0) duration. GA_1 (■), GA_3 (□) and 15β-OH-GA_3 (●) were applied to a single leaf at the doses (per plant) indicated.

Acknowledgments--The authors acknowledge research support from Operating and International Collaborative Research Grants of the Natural Sciences and Engineering Research Council of Canada (RPP). The able technical support of Mr. Bruce Twitchin is gratefully acknowledged, as is the gift of the sample of GA_{32} from Dr. N. Murofushi, Department of Agricultural Chemistry, University of Tokyo.

LITERATURE CITED

1. BERNIER G 1988 The control of floral evocation and morphogenesis. Annu Rev Plant Physiol 39: 175-219
2. BUBAN T, M FAUST 1982 Flower bud induction in apple trees: internal control and differentation. Hortic Rev 4: 174-203
3. DUNBERG A, G MALMBERG, T SASSA, RP PHARIS 1983 Metabolism of tritiated gibberellins A_4 and A_9 in Norway spruce, *Picea abies* (L.) Karst. Effects of a cultural treatment known to enhance flowering. Plant Physiol 71: 257-262
4. DUNBERG A, PC ODEN 1983 Gibberellins and conifers. *In* A Crozier, ed, The Biochemistry and Physiology of Gibberellins, Vol 2. Praeger, New York, pp 211-296
5. GOLDSCHMIDT EE, SP MONSELISE 1972 Hormonal control of flowering in citrus and some other woody perennials. *In* DJ Carr, ed, Plant Growth Substances. Springer-Verlag, Berlin, pp 758-765
6. HALEVY AH 1985 Handbook of Flowering, Vols I, II, III, IV. Boca Raton: CRC Press
7. HALEVY AH 1986 Handbook of Flowering, Vol. V. CRC Press, Boca Raton

8. HALEVY AH 1988 Handbook of Flowering, Vol VI. CRC Press, Boca Raton (in press)
9. KING RW, RP PHARIS, LN MANDER 1987 Gibberellins in relation to growth and flowering in *Pharbitis nil* Chois. Plant Physiol 84: 1126-31
10. LOONEY NE, RP PHARIS, M NOMA 1985 Promotion of flowering in apple trees with gibberellin A_4 and C-3 epi-gibberellin A_4. Planta 165: 292-294
11. MACMILLAN J, BO PHINNEY 1987 Biochemical genetics and the regulation of stem elongation by gibberellins. *In* DJ Cosgrove, DP Knievel, eds, Physiology of Cell Expansion During Plant Growth. The Am Soc Plant Physiologists, Bethesda, MD, USA, pp 156-171
12. PHARIS RP, LT EVANS, RW KING, LN MANDER 1987 Gibberellins, endogenous and applied, in relation to flower induction in the long-day plant *Lolium temulentum*. Plant Physiol 84: 1132-38
13. PHARIS RP, RW KING 1985 Gibberellins and reproductive development in seed plants. Annu Rev Plant Physiol 36: 517-68
14. PHARIS RP, CG KUO 1977 Physiology of gibberellins in conifers. Can J For Res 7: 299-325
15. PHARIS RP, NE LOONEY, LN MANDER 1986 Promotion of flowering in woody angiosperms. UK Patent Applic No 8502424 filed 31 Jan 1985, updated Jan 1986
16. PHARIS RP, SD ROSS 1986 Pinaceae hormonal promotion of flowering conifers. *In* AH Halevy, ed, Handbook of Flowering, Vol V. CRC Press, Boca Raton, FL pp 269-286
17. PHARIS RP, JE WEBBER, SD ROSS 1987 The promotion of flowering in forest trees by gibberellin $A_{4/7}$ and cultural treatments: a review of the possible mechanisms. For Ecol Manage 19: 65-84
18. ROSS SD, MP BOLLMAN, RP PHARIS, GB SWEET 1984 Gibberellin $A_{4/7}$ and the promotion of flowering in *Pinus radiata*: Effects of partitioning of photoassimilate within the bud during primordia differentiation. Plant Physiol 76: 326-330
19. ROSS SD, RP PHARIS, WD BINDER 1983 Growth regulators and conifers: their physiology and potential uses in forestry. *In* LG Nickell, ed, Plant Growth Regulating Chemicals. CRC Press, Boca Raton, FL, pp 35-78
20. ROSS SD, JE WEBBER, RP PHARIS, JN OWENS 1985 Interaction between gibberellin $A_{4/7}$ and rootpruning on the reproductive and vegetative process in Douglas-fir. I. Effects on flowering. Can J For Res 15: 341-347
21. SCHWABE WW, AH AL-DOORI 1973 Analysis of a juvenile-like condition affecting flowering in the black currant (*Ribes nigrum*). J Exp Bot 24: 696-81
22. VINCE-PRUE D, B THOMAS, KE COCKSHULL 1984 Light and the Flowering Process. Academic Press, London
23. WEBBER JE, SD ROSS, RP PHARIS, JN OWENS 1985 Interaction between gibberellin A_{4+7} and rootpruning on the reproductive and vegetative process in Douglas-fir. II. Effects on shoot growth. Can J For Res 15: 348-353
24. ZEEVAART JAD 1983 Gibberellins and flowering. *In* A Crozier, ed, The Biochemistry and Physiology of Gibberellins, Vol 2. Praeger, New-York, pp 333-74
25. ZIMMERMAN RH, WP HACKETT, RP PHARIS 1985 Hormonal aspects of phase change and precocious flowering. *In* RP Pharis, DM Reid, eds, Encyclopedia of Plant Physiology, New Series, Vol 11. Springer-Verlag, New York, pp 79-115

EVENTS OF THE FLORAL TRANSITION OF MERISTEMS

GEORGES BERNIER

*Département de Botanique, Université de Liège,
Sart Tilman, B-4000 Liège, Belgium*

The changeover from leaf to flower production at shoot meristems is a dramatic example of a morphogenetic switch, just like those occurring at various stages of animal embryogenesis.

The floral transition of meristems is usually investigated in photoperiodic species reacting to a single inductive cycle because they are the only species in which this transition occurs with sufficient synchrony among the individual plants of a population: this is illustrated in Figure 1 in the case of the LD plant *Sinapis alba* induced by a single LD. For induction (curve A), the earliest plants are only 6 h ahead of the slowest ones. For translocation of the leaf-generated floral stimulus (curve B) and initiation of the first flower (curve C), this time difference has increased to 12 and 24 h, respectively, indicating that the synchrony caused by the LD is progressively lost with time. Populations of plant species requiring several inductive cycles are very heterogeneous at all times because different individuals require, as a rule, different numbers of inductive cycles and are thus at various stages of the floral transition at any collection time after the start of induction.

My aim in this paper is to summarize recent work on the experimental system shown in Figure 1. The apical meristem of *Sinapis* is producing leaves at the start of the LD and continues to do so for the next 40 h. Its morphogenetic activities then cease for about 1 d and resume around 60 h after the start of the LD by the initiation of flower primordia. There is no intermediate appendage formed between the last leaf and the first flower. The timing of changes occurring during the transition will be expressed in hours after start of the LD.

It is generally believed that an irreversible commitment to form flowers, called floral evocation, is reached after some time during the floral transition (2, 3). I shall attempt here to distinguish between the events of the transition that are integral parts of evocation and those that are the consequences of it.

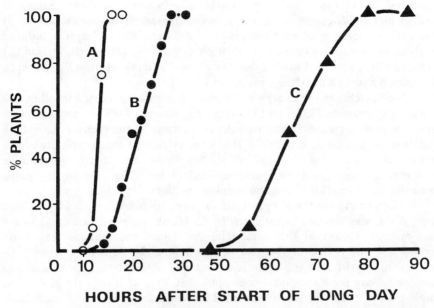

FIG. 1. Flowering response of *Sinapis* plants induced by a single LD. A, Induction curve obtained by exposing groups of plants to one LD of increasing lengths. B, Translocation curve for the floral stimulus obtained by removal of induced leaves at various times in plants exposed to one 22-h LD. C, Initiation curve obtained by exposing plants to one 22-h LD and determining at various times afterward the percent of plants having initiated their first flower (using histological sections of apical meristems).

CHANGES IN ENERGY METABOLISM

Meristematic tissues are nonphotosynthetic. Their energy is supplied essentially by more mature tissues, and energy transduction within the apex is by glycolysis and respiration. Since there is commonly an accelerated tempo of activities in the apex of plants induced to flower, one may expect changes in their energy metabolism.

In the apex of induced *Sinapis* plants there is indeed an increase in the sucrose level, followed by rises in acid invertase activity, ATP level, mitochondrion number per cell, succinic dehydrogenase activity, and starch level (summarized in 3 and 6). These changes occur quite early during the floral transition. In fact, the elevation in sucrose level is already detectable 10 h after the start of the inductive LD, actually 2 h after the start of the photoextension of this LD (7). Work with $^{14}CO_2$ indicates that there is no modification of the supply of the recently-synthesized assimilates for the apical bud that can account for the early increase in sucrose and other soluble sugars (8). Thus, an early remobilization of reserve carbohydrates stored within the apical bud itself, or in other plant parts, might have occurred, but this remains to be investigated. In any case, these changes in energy metabolism seem to be an essential component of evocation since (*i*) the early increase in sucrose level is observed even when *Sinapis* is induced without any extension of the period of photosynthetic activity (3, 7), and (*ii*) flowering is suppressed when 2,4-dinitrophenol is applied to the *Sinapis* apex between 8 and 40 h after the start of the LD (5).

Similar changes in energy metabolism have been reported in other LD plants, *e.g. Brassica* (23), and SD plants, *e.g.* tobacco (18). In the LD plant spinach, early increases in the activities of enzymes of the pentose phosphate pathway have been detected (1). However, in view of the complexity of the results in this species (11), further studies should use strictly quantitative microtechniques based on pyridine nucleotides (22) instead of semi-quantitative estimations based on staining reactions.

The situation in the SD plant *Xanthium strumarium*, induced by a single long night, was recently investigated by C. Mirolo (unpublished data) in my laboratory. Export of the floral stimulus out of the induced leaf starts between 4 and 8 h after the end of the long night, at a time when ^{14}C-assimilate import into the apex is decreased. Assimilate import starts to increase above the noninduced control level at 40 to 42 h after the end of the night and, at about the same time, acid invertase activity and mitochondrion number per cell also start to increase (see also 13). The reproductive structures (bracts) are not formed in this system before 100 h after the end of the night. The levels of soluble sugars and starch are never significantly different in apices of induced *versus* noninduced plants. Thus, despite the reduced duration of photosynthetic activity due to the long night and the early decrease in assimilate import, the sugar level in the apex is not decreased. Remobilization of storage carbohydrates is likely again. A brief light break given at the midpoint of the long night totally suppresses flowering, as well as the early drop in assimilate import and the rise in invertase activity, but not the late increase in assimilate import.

The available evidence supports the view that in *Xanthium*, like in *Sinapis*, some aspects of the energy metabolism are essential components of evocation.

CHANGES IN GENE EXPRESSION

Morphogenetic transitions are usually associated with specific changes in gene expression. The floral transition is no exception to this rule. Using two-dimensional (2D)-PAGE separation of ^{35}S-labeled polypeptides, it was found that the polypeptide complement of *Sinapis* meristems is changed before flower initiation starts (20). More recent work by F. Cremer (unpublished data) reveals that several quantitative and possibly few qualitative differences in the complement of recently-synthesized polypeptides are already detectable in meristems labeled from 24 to 27 h after the start of the inductive LD. So far, no analysis has been made before 24 h. Interestingly, these changes in protein synthesis occur at the time when ethionine, applied on the apex, is most inhibitory to flowering (16). This suggests that at least some of these changes are essential for evocation.

The protein complement of apices of *Silene coeli-rosa*, induced to flower by 7 LD, has also been analyzed by 2D-PAGE using mini-gels and silver staining (10). Qualitative differences in the number and location of polypeptides are first detected only on the 8th day after start of induction, *i.e.* 24 h before the sepals of the first flower are initiated. At that stage, about one-third of the polypeptides of transitional apices are different from those of vegetative apices. This change seems relatively late, but labeling of recently-synthesized polypeptides would have certainly revealed earlier changes.

Because of the minute amounts of tissue available so far, there is no study of the mRNA complement of shoot meristems. However, if the inhibitor work is any guide, it may be tentatively concluded that some changes in RNA synthesis in the *Sinapis* apex during the time interval from 8 to 24 h are essential since this is the period when 5-fluorouracil is most inhibitory to flowering (16).

The role, if any, of the changing proteins in evocation is entirely unknown. Further progress will require identification of some of these proteins or of their mRNA. This may not be easy given that available antibodies, or cDNA probes, are essentially from mature organs and that the proteins and mRNA involved in evocation are presumably specific to meristems and, thus, not present in mature tissues.

CHANGES IN CELL PROLIFERATION AND CHROMATIN STRUCTURE

The vegetative meristem of *Sinapis* is a mosaic of rapidly-cycling and slowly-cycling cells called cycling and noncycling cells, respectively. The first change after floral induction is a mitotic wave peaking at 26 to 30 h after the start of the LD, resulting from both a shortening of the G_2 phase of cycling cells and a return to cycling of noncycling G_2 cells (12). This is followed by a shortening of G_1 and S in cycling cells, resulting in a wave of cells duplicating DNA at 38 h. The shortening of S was essentially attributable to the halving of DNA replicon size (Fig. 2; 15). A second mitotic wave occurs later, coincident with the initiation of the first flower at about 60 h, and since the time interval separating the two mitotic waves is identical to the shortened cell cycle length, it is clear that cell division is synchronized during the floral transition. Part of these changes is caused by an increase in the cytokinin supply from the leaves (4, 17).

Similar changes in cell proliferation are observed in several other species. For example, both the decrease in cell cycle length and synchrony of cell division occur in transitional meristems of the LD plant *Silene coeli-rosa* (9) and *Lolium temulentum* (21 and unpublished data of A. Jacqmard and J.C. Ormrod).

Vegetative

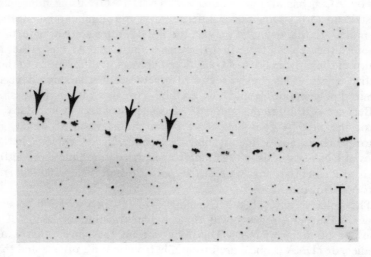

Evoked

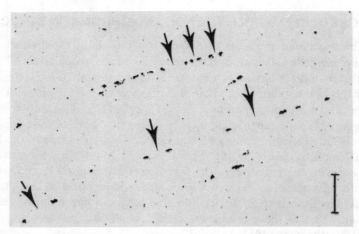

FIG. 2. Autoradiographs of DNA fibers from cells of vegetative and evoked meristems of *Sinapis* labeled with [^{3}H]TdR for 4 h. The arrows indicate midpoints of pairs of labeled DNA segments. Bar, 20 μm. Reproduced from Jacqmard and Houssa (15) with permission from Academic Press.

Five-fluorodeoxyuridine, applied to the *Sinapis* apex at the time of the wave of DNA synthesis, suppresses flowering (16). Thus, the stimulation of cell proliferation in this species seems to be another essential component of evocation.

Scanning cytophotometry and computer analysis show that the distribution of nuclear DNA between the dispersed and condensed fractions of chromatin is changed in *Sinapis* at the floral transition (14). In both G_1 and G_2 nuclei, there is an increase in the proportion of DNA present in the dispersed fraction and, thus, a decrease in the condensed fraction. This change is observed from 30 h after the start of the LD in G_1 nuclei, but only 20 h later in G_2 nuclei. This decondensation of chromatin is presumably related to the change in replicon length noted previously. Changes in chromatin structure might also be related to the activation and deactivation of many genes, *i.e.* the changes in gene expression, that were shown above to occur in transitional apices.

CHANGES IN GROWTH PARAMETERS

The apical growth pattern is completely altered during the floral transition in *Sinapis*. The rate of appendage (leaves or flowers) production is increased from 24 h after start of the LD, but this increase is statistically significant only at 72 and 120 h when flower primordia are being initiated (Table I). During the 5-d period investigated, the plastochron duration

Table I. *Appendage production by the apex of Sinapis plants kept in SD (vegetative) or induced to flower by one LD*[a]

Hours After Start of LD	Appendage (Leaves + Flowers) Number		Increment of Appendage Number Per Day (Relative to 0 Time)	
	SD	LD	SD	LD
0	28.0 (0.7)		--	
24	28.2 (0.6)	29.3 (0.4)	0.2	1.3
48	28.4 (0.7)	29.7 (0.6)	0.2	0.9
72	29.6 (0.6)	34.6* (0.8)	0.5	2.2
120	30.8 (0.8)	37.6* (1.2)	0.6	1.9

[a] Standard errors are in parentheses. Values marked * are significantly different for the two treatments (SD, LD) at the 95% level of the Student's *t* Test.

decreases from about 2.3 d to 8.4 h. Size and shape of the meristem are also changed: width and height start to increase at 46 h after the start of the LD, whereas doming (increase in height relative to width) occurs at 54 h (data not shown). At 48 h, there is a significant increase in length of the uppermost embryonic internodes and a progressively more acropetal location of axillary bud primordia (Bernier, unpublished data). At 72 h, the plastochron ratio, *i.e.* the ratio of successive radii of leaf primordia, starts to decrease. The longitudinal component of apical growth (length of embryonic internode) is clearly affected before, and more markedly than, the radial component (plastochron ratio).

Also, it should be pointed out that most changes of growth parameters are very late compared to the molecular and cellular changes discussed above.

Again, similar changes in apical growth have been found in many other species (2, 3, 19), but in these cases the timing relative to other events of the floral transition is poorly known.

CONCLUDING REMARKS

As far as the available evidence goes, more or less similar changes seem to occur in several unrelated species during the floral transition. Thus, I tentatively conclude that many features of this transition are fairly universal, even though some other details may vary considerably between different species.

As discussed elsewhere (3, 5), various metabolic inhibitors can prevent flowering in *Sinapis* when applied directly to the apex. Each compound acts during a particular time interval relative to the start of the inductive LD, but none is effective after about the 44th h. On this basis, it has been proposed that the irreversible commitment to flower morphogenesis, *i.e.* the completion of evocation, occurs at about 44 h, more than a half day before initiation of the first flower (2, 3, 16). The events of evocation are thus those occurring before 44 h and which are essential for a successful floral transition. As seen above, they all occur at the molecular and cellular levels. The earliest changes at the higher levels of organization, *i.e.* changes in apical growth (morphology), start just after the 44th h and, thus, are viewed as the outcome of evocation and the first steps toward flower morphogenesis.

Now, if the decondensation of chromatin is accepted as a marker of evocation, it may be tentatively concluded that the meristem of *Sinapis* consists, from 30 to 50 h after start of the LD, of a mosaic of evoked and nonevoked cells. Evocation would then be a progressive process, and its completion is possibly reached only when a sufficient proportion of cells is evoked.

The nature of evocation is not entirely clear. A sustained increase in cell proliferation requires larger expenditures and turnover of energy-supplying compounds, and thus will depend at some stage upon an increase in energy metabolism. Recent results of C. Houssa (unpublished data) in my laboratory indicate that this is indeed what happens in *Sinapis* (2). The mitotic activation is, in turn, responsible for the increased meristem size which apparently results in the elimination of the pattern of cellulose reinforcement at the dome center (P. B. Green, this volume, pp. 58-75). This latter change seems critical since it probably permits the appendage arrangement typical of the inflorescence (or flower) to replace the arrangement characteristic of the leafy shoot.

The significance of other evocational events, especially the early changes in gene expression, however, remains elusive (2).

Acknowledgments--This work is supported by research grants from the Belgian Government and University of Li ge (Action de Recherche Concert e 88/93-129), FRFC (Grants 2.4507.87 and 2.9009.87) and IRSIA.

LITERATURE CITED

1. AUDERSET G, PB GAHAN, AL DAWSON, H GREPPIN 1980 Glucose-6-phosphate dehydrogenase as an early marker of floral induction in shoot apices of *Spinacia oleracea* var. Nobel. Plant Sci Lett 20: 109-113
2. BERNIER G 1988 The control of floral evocation and morphogenesis. Annu Rev Plant Physiol Plant Mol Biol 39: 175-219
3. BERNIER G, JM KINET, RM SACHS 1981 The Physiology of Flowering, Vol II. CRC Press, Boca Raton, FL
4. BERNIER G, P LEJEUNE, A JACQMARD, JM KINET 1989 Cytokinins in flower initiation. *In* RP Pharis, SB Rood, eds, Plant Growth Substances 1988. Springer Verlag, Berlin (in press)
5. BODSON M 1985 *Sinapis alba*. *In* AH Halevy, ed, Handbook of Flowering, Vol. IV. CRC Press, Boca Raton, FL, pp 336-354
6. BODSON M, G BERNIER 1985 Is flowering controlled by the assimilate level? Physiol V g 23: 491-501
7. BODSON M, WH OUTLAW JR 1985 Elevation in the sucrose content of the shoot apical meristem of *Sinapis alba* at floral evocation. Plant Physiol 79: 420-424
8. BODSON M, B REMACLE 1987 Distribution of assimilates from various source-leaves during the floral transition of *Sinapis alba* L. *In* JG Atherton, ed, The Manipulation of Flowering. Butterworths, London, pp 341-350
9. FRANCIS D, RF LYNDON 1985 The control of the cell cycle in relation to floral induction. *In* JA Bryant, D Francis, eds, The Cell Division Cycle in Plants. Cambridge Univ Press, Cambridge, pp 199-215
10. FRANCIS D, J REMBUR, A NOUGAREDE 1988 Changes in polypeptide composition in the shoot apex of *Silene coeli-rosa* during floral induction. C R Acad Sci Paris Vol 307, Serie III, pp 763-770
11. GAHAN PB, G AUDERSET, DF DARMIGNAC, H GREPPIN 1987 The relationship between the activities of the pentose phosphate pathway and glycolysis during early stages of floral induction in spinach. Histochemistry 87: 289-291

12. GONTHIER R, A JACQMARD, G BERNIER 1987 Changes in cell cycle duration and growth fraction in the shoot meristem of *Sinapis* during floral transition. Planta 170: 55-59
13. HAVELANGE A 1980 The quantitative ultrastructure of the meristematic cells of *Xanthium strumarium* during the transition to flowering. Am J Bot 67: 1171-1178
14. HAVELANGE A, JC JEANNY 1984 Changes in density of chromatin in the meristematic cells of *Sinapis alba* during transition to flowering. Protoplasma 122: 222-232
15. JACQMARD A, C HOUSSA 1988 DNA fiber replication during a morphogenetic switch in the shoot meristematic cells of a higher plant. Exp Cell Res 179: 454-461
16. KINET JM, M BODSON, AM ALVINIA, G BERNIER 1971 The inhibition of flowering in *Sinapis alba* after the arrival of the floral stimulus at the meristem. Z Pflanzenphysiol 66: 49-63
17. LEJEUNE P, JM KINET, G BERNIER 1988 Cytokinin fluxes during floral induction in the long day plant *Sinapis alba* L. Plant Physiol 86: 1095-1098
18. LU CC, JF THOMAS 1988 Succinate dehydrogenase activity in the shoot apex of tobacco during floral initiation. Plant Physiol 86 (Suppl): 148
19. LYNDON RF, NH BATTEY 1985 The growth of the shoot apical meristem during flower initiation. Biol Plant 27: 339-349
20. LYNDON RF, A JACQMARD, G BERNIER 1983 Changes in protein composition of the shoot meristem during floral evocation in *Sinapis alba*. Physiol Plant 59: 476-480
21. ORMROD JC, BERNIER G 1989 Cell cycle patterns in the shoot apex of *Lolium temulentum* cv Ceres during the transition to flowering following a single long day. (submitted)
22. OUTLAW WH JR 1980 A descriptive evaluation of quantitative histochemical methods based on pyridine nucleotides. Annu Rev Plant Physiol 31: 299-311
23. PETERSEN K, AR ORR 1983 Histochemical study of enzyme activity in the shoot apical meristem of *Brassica campestris* L. during transition to flowering. I. Succinic dehydrogenase. Bot Gaz 144: 338-341

FLORAL INITIATION AS A DEVELOPMENTAL PROCESS

Carl N. McDaniel

*Plant Science Group, Department of Biology,
Rensselaer Polytechnic Institute, Troy, NY 12180, USA*

Flowering plants have evolved over the past 100 million years to give rise to more than 200,000 species which have exploited every habitat except for the deep ocean waters. All aspects of the flowering process have evolved such that in each species specific control mechanisms have emerged to permit reproductive success. It is not surprising that the investigation of flowering across all of this diversity has generated an enormously complex array of data which has bewildered us and defied simple explanations (3, 9, 23).

A formidable body of experimental evidence has firmly established that the leaves play a major role in the regulation of floral initiation in many plants (10). Although the stimulatory role played by the leaves can most easily be explained by invoking a hormone, florigen, no one has been able to identify florigen. Bernier (2) has stated that the classical model for the control of floral initiation, which gives florigen the major role, is too simplistic to do justice to the complexity of the observations reported. Bernier has convincingly argued that a host of processes throughout the plant are intimately related to the mechanism(s) controlling floral initiation. However, I do not believe that the reality of these various processes negates the validity of the classical view. It has been my contention that our enormous wealth of information about flowering in a wide diversity of angiosperms has confused the issue because we are trying to fit millions of years of evolution into one universal pattern of floral initiation as opposed to seeking a basic sequence of developmental processes which have been modified during the evolution of each species (12). I believe we need to do at least three things in order to make fundamental progress in understanding floral initiation: *i*) appreciate and experimentally exploit the genetic diversity which has arisen during the evolution of angiosperms, *ii*) adopt a more developmentally oriented view of floral initiation, and *iii*) establish, in a group of representative species, which developmental events are controlled. This information will permit us to piece together the global pattern of the processes which control initiation and, at the same time, elucidate the variations in the basic pattern which have evolved.

Here I will present what we have learned about the initiation of flowering in *Nicotiana* and cast our observations in a developmental perspective.

MATERIAL AND METHODS

We have employed two dayneutral (DN) varieties, *Nicotiana tabacum* cv Wisconsin 38 and cv Hicks; two SD varieties, *N. tabacum* cv Maryland Mammoth and cv Hicks Maryland Mammoth; and one LD species, *N. sylvestris*. Plants were grown in growth rooms, in greenhouses, and in the field under various conditions. Procedures for establishing floral determination and for assessing the roles of leaves and roots in floral initiation have been presented in other places (7, 11, 19, 20).

RESULTS AND DISCUSSION

Inductive Capacity of Leaves. Under a given set of environmental conditions the terminal meristem of a DN (18, 22) or a photoperiodic tobacco plant (7, 15) produces a precisely regulated number of nodes and then forms a flower. Only one to several leaves are necessary to induce the terminal meristem to flower (10), and it has been shown for DN tobacco that their location on the main axis is not critical (4, 11). Only four apical or basal leaves of at least 10 cm in length are required for the terminal meristems of DN tobaccos to produce the same number of nodes as meristems of plants which have all of their leaves (6, 11). The inductive capacity of apical and basal leaves in SD Maryland Mammoth tobaccos is different. Four basal leaves which are at least 10 cm in length are not capable of inducing the meristem to flower while as few as two, 10-cm, apical leaves will induce flowering after the meristem has produced the same number of nodes as untreated plants (6). This difference in inductive capacity of basal leaves is seen in two cv of *N. tabacum*, Hicks and Hicks Maryland Mammoth, which are isogenic except for the Maryland Mammoth allele.

Root Influence. When the apical portion of a DN *N. tabacum* cv Wisconsin 38 plant is rooted each time five to seven leaves at least 10 cm in length are present, the meristem continues to produce leaves and does not flower (11). SD *N. tabacum* cv Maryland Mammoth behaves differently. After producing up to 10 more nodes than untreated plants, the terminal meristem flowers despite the continual rooting (6).

Floral Determination in Apical Buds. Rooting assays have established for *N. tabacum* cv Wisconsin 38 (19), *N. tabacum* cv Maryland Mammoth (7), and *N. sylvestris* (15) that the terminal apical bud becomes determined for floral development prior to the initiation of the terminal flower and that one to three leaf primordia are initiated after a bud becomes florally determined.

Floral Determination in Internode Cells. It has been known for a long time that internode cells from the inflorescence of flowering DN tobaccos have a high capacity to form floral shoots when cultured (1). We have observed that internode cells from young vegetative plants form a low frequency of floral shoots in culture (14) and this frequency increases substantially beginning at about the time that the terminal bud becomes florally determined (20). The frequency of floral shoots increases from about 0.1% in internodes of young vegetative plants to about 75% in the inflorescence internodes of plants at anthesis of the terminal flower.

Analysis of Results. The above observations are summarized in Figure 1. The leaves of the tobacco plant send an inductive signal which can act both on organized meristems and on other cells in the plant to commit them to a floral state. Thus, the inductive signal acts on competent cells to change their developmental state. The effectiveness of the inductive signal sent by the leaves is a function both of genotype and of environment. As has long been appreciated, the Maryland Mammoth allele usually prevents the sending of an effective signal under noninductive (LD) conditions; in addition, this gene modulates the effectiveness of the leaf signal under inductive (SD) conditions. That is, the most basal leaves of Maryland Mammoth plants are not inductively active, while the apical leaves send a very effective signal.

Leaf-signal effectiveness in *N. tabacum* cv Wisconsin 38 plants grown under LD conditions contrasts with that of the Maryland Mammoth types. In Wisconsin 38, the two basal-most leaves of at least 10 cm in length can induce the meristem into a floral state while two apical leaves of the same length are not inductively active. The inductive activity of basal leaves on *N. tabacum* cv Wisconsin 38 plants is also indicated by the fact that cultured stem tissues from young vegetative plants form some *de novo* floral shoots.

The effectiveness of the inductive signal from the leaves can be modified by an input from the roots. This is most clearly seen in DN *N. tabacum* cv Wisconsin 38 where the terminal meristem does not flower unless it is separated from the roots by some minimal number of nodes. Thus, there seems to be an interplay between leaf and root inputs which involves the number of nodes separating the terminal meristem from the roots. Since Maryland Mammoth types cannot be kept vegetative by continual rooting, the inductive leaf signal produced by leaves which are initiated later in development appears to override the root signal or node requirement.

Developmental Black Box. Floral initiation and subsequent floral morphogenesis can be analyzed from a developmental perspective. We have depicted the events and processes which occur during the making of an individual in a simple but versatile diagram (16; Fig. 2). Beginning with a zygote, cell divisions produce a complex array of cell types which are arranged into highly ordered spatial patterns such that for each species recognizable morphologies appear. How is this accomplished? Mostly, we do not know. However, to discuss developmental phenomena, we often employ the terms depicted in the "Developmental black box." The terms can refer to a cell, a group of cells, a tissue, or an organ. For simplicity, I will talk at the level of cells, but bear in mind that the phenomena described may be relevant at many levels of organization and, in most instances, we do not know which level is applicable.

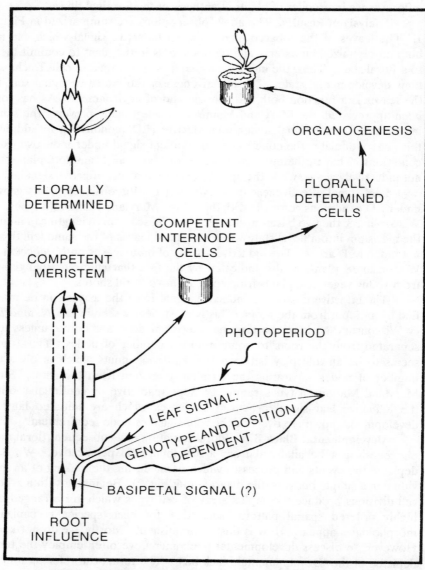

FIG. 1. Inputs which regulate floral determination in the genus *Nicotiana*. See text for explanation and discussion.

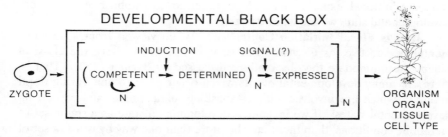

FIG. 2. Developmental processes and sequences which occur as a zygote becomes a mature organism. See text for explanation and discussion. Reproduced from McDaniel et al. (16) with permission from Butterworths.

Cells are competent if they are capable of responding to a developmental signal in a specific, predictable way. In many documented cases the signal is an inductive signal and, in response, the cells become determined for a new or more restricted developmental fate (8, 13, 17, 21). The competence for a specific response can change, and it has been observed that cells may be competent for a specific response for only a limited time period. The competent-induction-determined sequence may be repeated and this allows cells to revert to an earlier state of determination, as well as to transit to new determined states. Currently, we do not know the molecular basis for any state of competence or for any state of determination; however, these states must reflect some change, and it is assumed that ultimately these differences will be known. Following induction, some determined states may be expressed such that they can be measured by current techniques. For example, the cells may change morphology, synthesize a new protein, or gain the capacity to respond to a hormone. In other cases, expression may not directly follow induction but require an additional developmental signal. Before cells reach their ultimate developmental fate, the sequence of events from competence to expression may be repeated, thereby allowing for a return to an earlier condition or for the expression of new states or conditions. The end result of a vast multitude of these developmental sequences is an organism.

Our laboratory has attempted to analyze floral initiation in tobacco as a developmental process. We have succeeded in identifying one stable developmental state as exemplified by buds or *de novo* buds whose meristems produce a terminal flower after initiating a limited number of nodes; this developmental behavior contrasts to that of other buds which initiate many nodes before making a terminal flower. We have called this state "floral determination" because the assay we employed assessed developmental fate in several environments, the classical operation performed to establish the existence of a determined state (13, 21). However, it is possible that we have been characterizing a competence state. The isolation of the inductive signal and/or additional experiments will be required to establish exactly what developmental state we are characterizing.

How is Floral Initiation Controlled? When viewed from a developmental perspective, as discussed above, it is possible to have the classical scheme accommodate the observations discussed by Bernier (2, 3). That is, there are many developmental events which occur prior to and during floral initiation which are involved with establishing, maintaining, and expressing developmental states. If we imagine a developmental state as a stable set of metabolic conditions, then there may be more than one way by which a set of metabolic conditions can be established (5). Thus, it may be possible to short circuit a normal sequence of events by employing an alternative route which may or may not occur in nature. Depending upon the relationships among different groups of cells and their developmental states, sequential ordering of developmental states may or may not be critical. Thus, the classical model, which ascribes a major role to the leaf inductive signal in floral initiation, may describe the final events in the normal sequence of events which lead to the onset of floral morphogenesis. Bernier's scheme of parallel, interacting sequences of events may describe some of the many developmental sequences that normally lead up to the inductive activity of the leaf signal which then brings about a state of floral determination in a shoot apical meristem.

Acknowledgments--I am indebted to Kelly Dennin, Joan Gebhardt, Karla Sangrey, Susan Singer, Holly Vogel, and Maura Westerdahl for their research efforts and many discussions over the years which made this manuscript possible and Susan Smith for her critical review of the manuscript. This work was supported by grants from the National Science Foundation (DCB 87-09871) and from the U.S. Department of Agriculture Competitive Grants Program (86-CR CR-1-2053).

LITERATURE CITED

1. AGHION-PRAT D 1965 Neoformation de fleurs *in vitro* chez *Nicotiana tabacum* L. Physiol Veg 3: 229-303
2. BERNIER G 1986 The flowering process as an example of plastic development. *In* DH Jennings, AT Trewavas, eds, Plasticity in Plants. Company of Biologists, Cambridge, pp 257-286
3. BERNIER G 1988 The control of floral evocation and morphogenesis. Annu Rev Plant Physiol Plant Mol Biol 39: 175-219
4. CHAILAKHYAN MKH, KHK KHAZHAKYAN 1974 The interaction of leaves and shoots in the flowering of photoperiodically neutral species. Dokl Akad Nauk SSSR 217: 1214-1217
5. CHRISTIANSON ML 1985 An embryogenic culture of soybean: Towards a general theory of somatic embryogenesis. *In* RR Henke, KW Hughes, MP Constantin, A Hollaender, eds, Tissue Culture in Forestry and Agriculture. Plenum, NY, pp 83-103

6. GEBHARDT JS 1987 Induction and determination for floral development in short-day *Nicotiana tabacum* L. MS Thesis, Rensselaer Polytechnic Institute, Troy, NY
7. GEBHARDT JS, CN MCDANIEL 1987 Induction and floral determination in the terminal bud of *Nicotiana tabacum* L. cv Maryland Mammoth, a short-day plant. Planta 172: 526-530
8. GURDON JB 1985 Introductory Comments. *In* J Sambrook, ed, Cold Spring Harbor Symp on Quant Biol, Vol L, Molecular Biology of Development. Cold Spring Harbor Laboratory, Cold Spring Harbor, NY, pp 1-10
9. LANG A 1965 Physiology of flower initiation. *In* W Ruhland, ed, Encyclopedia of Plant Physiology, Vol 15 (Part 1). Springer-Verlag, Berlin, pp 1380-1536
10. LANG A 1987 Perspectives in flowering research. *In* JL Key, L McIntosh, eds, Plant Gene Systems and Their Biology, UCLA Symposia. Alan R Liss, NY, pp 3-24
11. MCDANIEL CN 1980 Influence of leaves and roots on meristem development in *Nicotiana tabacum* L. cv Wisconsin 38. Planta 148: 462-467
12. MCDANIEL CN 1984 Shoot meristem development. *In* DJ Carr, PW Barlow, eds, Positional Controls in Plant Development. Cambridge University Press, Cambridge, pp 319-347
13. MCDANIEL CN 1984 Competence, determination and induction in plant development. *In* G Malacinski, ed, Pattern Formation: A Primer in Developmental Biology. Macmillan, NY, pp 393-412
14. MCDANIEL CN, KA SANGREY, DE JEGLA 1989 Cryptic floral dedertmination: explants from vegetative tobacco plants have the capacity to form floral shoots *de novo*. Submitted to Dev Biol
15. MCDANIEL CN, SR SINGER, KA DENNIN, JS GEBHARDT 1985 Floral determination: Timing, stability, and root influence. *In* M Freeling, ed, Plant Genetics. Alan R Liss, NY, pp 73-87
16. MCDANIEL CN, SR SINGER, JS GEBHARDT, KA DENNIN 1987 Floral determination: A critical process in meristem ontogeny. *In* JG Atherton, ed, The Manipulation of Flowering. Butterworths, London, pp 109-120
17. NIEUWKOOP PD 1985 Inductive interaction and determination; A new approach to an old problem. *In* GM Edelman, ed, Molecular Determinants of Animal Form. Alan R Liss, NY, pp 59-71
18. SELTMAN H 1974 Effect of light periods and temperatures on plant form of *Nicotiana tabacum* L. cv Hicks. Bot Gaz 135: 196-200
19. SINGER SR, CN MCDANIEL 1986 Floral determination in the terminal and axillary buds of *Nicotiana tabacum* L. Dev Biol 118: 587-592
20. SINGER SR, CN MCDANIEL 1987 Floral determination in internode tissues of dayneutral tobacco first occurs many nodes below the apex. Proc Natl Acad Sci USA 84: 2790-2792
21. SLACK JMV 1983 From Egg to Embryo, Determinative Events in Early Development. Cambridge University Press, Cambridge
22. THOMAS TF, CE ANDERSON, CD RAPER JR, RJ DOWNS 1975 Time of floral initiation in tobacco as a function of temperature and photoperiod. Can J Bot 53: 1400-1410
23. ZEEVAART JAD 1976 Physiology of flower formation. Annu Rev Plant Physiol 27: 321-348

SHOOT MORPHOGENESIS, VEGETATIVE THROUGH FLORAL, FROM A BIOPHYSICAL PERSPECTIVE

PAUL B. GREEN

*Department of Biological Sciences, Stanford University,
Stanford, CA 94305, USA*

The typical angiosperm shoot has a vegetative growth period followed by a floral stage. Each phase has cyclic development. There is usually a transduction mechanism to link environmental change to the transition between phases. We will not be concerned here with the transduction, but rather with the largely repetitive nature of the responding system, the apical meristem, and how it can shift from one cyclic pattern to another.

Vegetative and floral development are readily distinguished. Differences in biosynthetic activity are obvious (*e.g.* petal pigments are made only in flowers). They certainly relate to the numerous differences in gene expression that have been demonstrated (4). Differences in geometry are equally obvious. A flower is greatly modified from its presumed homolog, a vegetative branch. At the level of symmetry, one notes that the trimerous flowers of an iris arise on a stem with distichous leaf arrangement (zig-zag in a plane). The five-petaled, star-like flowers of the jade plant (*Crassula argentea*) arise on shoot structures with strict decussate phyllotaxis (4-fold symmetry). The distinctive geometrical changes, which are inherited as rigorously as the biosynthetic ones, are the subject of this paper.

The synthetic and geometrical aspects of development are species specific. The frame of reference to connect the genome to the biosynthetic changes is the best established causal sequence in biology. How to modify and extend this dogma to connect the genome to geometrical changes, beyond those of self-assembly, is not yet clear. There are two candidates for the appropriate conceptual supplement; both rely on the laws of physics.

The first proposal assumes that geometrical change is closely tied to change in concurrent specific synthesis, usually viewed as differential gene expression. The suggestion is that, to account for geometrical change, the genome first produces the components of a reaction-diffusion system. A pair of compounds called morphogens is capable of generating, reproducibly, progressive patterns of their own concentration (13, 18). Once distinctive chemical profiles are established, unique reactions occur at specific places, *e.g.* concentration peaks. These reactions could activate genes, hence, leaves would arise "where the leaf genes are turned on" (14). The reaction-diffusion process could operate repeatedly to subdivide space into ever smaller unique regions, down to the level of single cells. Combinatorial use of the genome could, in principle, make each cell of an organ different from the others. These concepts are clearly powerful in generating pattern. The main shortcoming of this explanatory scheme is that it does not deal explicitly with how chemical profile, and presumed synthetic activation, is converted to large-scale structure. Thus, Meinhardt (18) states, "The change of shape and form (the morphogenesis proper) which is thought to be a consequence of these primary differences, is not considered." Hence, one concern is that the connection of chemical pattern to physical form is not addressed. Another is that the morphogens have not been detected.

A second proposal considers construction directly; it has the cell as the central functional element (7, 19). On the one hand, the cell, with a specific set of developmental responses to chemical and physical input, is the product of the genome. On the other, these cell responses, when integrated over space and time, produce the three-dimensional structure. Because the set of responses can be constant, different ones being called forth at different times, the role of concurrent change in gene activation is diminished. Diverse structures can arise from minor variations on a set code of cell responses. In this scheme, there would be a group of consistent "shoot genes" encoding cell responses at the apex. The same set of cell responses would produce both leaf and stem primordia, these being comprised of similar cells. The leaf-specific genes would be turned on only later, in response to some unique feature tied to the physical establishment of the primordium. For example, its tip region would have a different gaseous environment and would synthesize IAA. In this scheme, organ-specific syntheses generally follow construction rather than precede it. The two functions of patterning and construction are tightly coupled, rather than being separated as in the first scheme. This chapter will analyze vegetative and floral development from this second perspective.

BIOPHYSICAL PRINCIPLES APPLIED TO THE SHOOT

The Shoot as a Developmental Engine

If one wishes to connect the shoot, a complex phenotype, to the genome, there are major simplifications that can be made immediately. Because of the repetitive nature of most shoot structures, the tip of the shoot can be viewed as a developmental "engine." The fundamental similarity to a mechanical engine is that the same geometrical configuration is attained repeatedly. One very major difference is that these identical configurations are comprised of different cells at each cycle. Another closely related difference is that the product of the engine is not torque, but rather is new volume in the form of a main axis and its appendages. Clearly, the genome produces the shoot by producing the engine that generates it.

The activity of a decussate meristem, as in maple or *Kalanchoe*, can be readily condensed to the engine format (Fig. 1). During each cycle, the elliptical apical dome must produce: a leaf pair, a stem segment, and a new dome ("itself") at 90° to the previous orientation. This is obviously a three-dimensional process and, at first glance, is relatively intractable for further analysis. The essential features can be retained, however, after the elements have been translated into two-dimensional phenomenology. The pertinent plane is not the traditional longitudinal section, but rather is the surface plane of the shoot tip. The ultimate link to the one-dimensional genome is that cytoskeletal responses, which are essential to histogenesis in the surface plane, are undoubtedly based on protein properties.

The Engine in Two Dimensions

The reduction of shoot structure to two dimensions starts with the fact that a plant organ can be approximated as an elongating cylinder (flattened in leaves and petals). The basis for elongation is that the cells in the meristem are transversely reinforced in their side walls by "hoops" of cellulose microfibrils. Elongation occurs at right angles to reinforcement. Because stresses accumulate at the organ surface (16), reinforcement in the outer epidermal wall is especially important. Hence, one can approximate an established organ as a single hoop-reinforced cylinder, the cylinder being a mosaic of outer epidermal walls (Fig. 1). By the same token, the incipient primordium is viewed as a hoop-reinforced low mound, as a system of concentric circles or rings (Fig. 2). When the concentric rings later protrude out of the plane, they become adjacent hoops in the new axis. This is true for

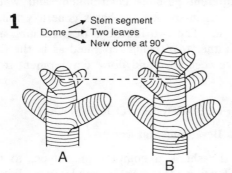
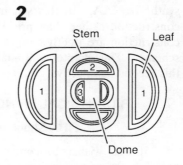

FIG. 1. The beginning (A) and the end (B) of a decussate vegetative cycle. The apical dome in B has produced a stem segment, two leaves, plus a new dome. Lines show hoop-reinforcement on the main axis and appendages.

FIG. 2. A decussate shoot in two dimensions, seen from above. Large hoops represent the cellulose alignment for the stem; smaller hoop regions are appendages (numbered in sequence of origin).

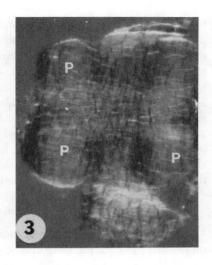

FIG. 3. Structure of flowering apices of *Kalanchoe*. Isolated surface cells from a developing flower, seen in polarized light with compensator rotated. Regions darker than the background have cellulose alignment parallel to the long axis of the page. Lighter regions have alignment normal to the dark regions. The lower central pair of dark regions, with a slightly lighter region in between, is the site for a stamen. Compare with Fig. 12a. X 200. From Dr. Amy Nelson.

the main axis (large hoops) and appendages (small hoops). Thus, the product of the developmental engine, a new axis, can be reduced to a concentric ring arrangement on the surface plane of the shoot. The problem of producing new three-dimensional axes is thereby reduced to the problem of generating specific new concentric ring-patterns on the meristem surface. The components of reinforcement patterns are deduced from polarized light images of isolated surface layers, as in Figure 3.

New Hoops Arise Within Old Ones

Given that the production of large-scale hoop reinforcement is the biophysical key to axis formation, there are two broad alternatives for shoot development. The apical dome could be relatively unstructured with regard to reinforcement. In this case, concurrent but independent processes would give rise to *de novo* hoop reinforcement on both the main axis and the appendages. Or, the dome itself could be hoop reinforced so the new ring patterns for appendages would be <u>intercalated</u>, *i.e.* partly derived from the pre-existing pattern of the main axis. In the cases studied so far, the latter has been seen for cellulose alignment. This alignment is thought to be based on microtubule orientation (12). The corresponding alignment for microtubules, though less clear, has been shown for the dome of *Vinca* (20, 21).

Old Organs Influence the Position of New Ones

It is a major assumption in shoot development that recently formed appendages influence the dome in cyclic fashion to perpetuate organ pattern. In the present context, this influence must result in new ring-reinforced areas arising locally inside an established one on the dome. In general, there is an acropetal (inward) propagation of large hoop structure on the dome. Systematic interruption of the continuous large-scale hoop pattern characterizes vegetative and reproductive development. A decussate vegetative pattern is shown in Figure 4 and a tetramerous floral pattern in Figure 5. New primordia arise at regions which are distant from the dome center and have sharp reinforcement curvature (x in Figs. 4 and 5). New rings arise through a 90° reinforcement shift which, after alignment is smoothed, establish a new appendage. Growth of these laterals then reshapes the dome so that the next set of sites, which are highly curved and distant from the center, lies at an angle to the previous one. In whorled structures, the angle is half that between appendages in the whorl, hence, the lateral organs "nest." The 90° shift in reinforcement, an axiality shift, is of great importance. It is believed to be a cell response, based on cytoskeletal behavior.

Changing Reinforcement Alignment

The microfibrillar reinforcement of a cylindrical cell is apparently based on the alignment of the microtubules in its cortical array (12, 17). The array

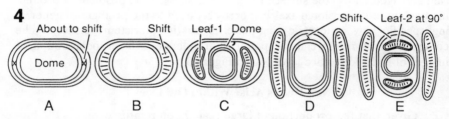

FIG. 4. Sequence in a vegetative decussate shoot. Lines show reinforcement; x shows the site of an imminent local 90° shift in reinforcement, an axiality shift. The site has high curvature, is distant from the center. Leaves formed by the shift extend the dome in D. New shifts in D make a second pair of leaves in E. They will stretch the dome to complete the cycle back to A.

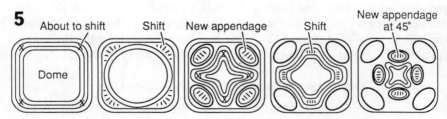

FIG. 5. Highly schematic flower development. Large-scale hoop reinforcement is propagated inwardly. Periodic axiality shifts (x) occur at the sharply curved corners of the dome, initiating new laterals, as above. Growth of laterals causes corners of the new dome to be offset, by 45°, from the previous stage, hence, organs "nest."

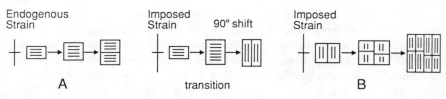

FIG. 6. Cell rules. Division and reinforcement are normal to the strain direction when it is moderate in rate and/or directionality (A). Under strong directional strain, division and reinforcement become parallel to the strain maximum (transition). As long as the imposed strain pattern persists, the cells continue to extend in the direction of stretch (B).

generally is transverse, *i.e.* parallel to the cellulose. The alignment is reestablished after the cell divides transversely. This is state A in Figure 6. In meristems, the shifts in alignment of cellulose reinforcement occur after cell division (5). Pairs of cells, recently formed by a longitudinal division, usually show cellulose alignment parallel to the new cross wall and to the long axis of the just-formed cells. This is the transition shown in Figure 6 (middle). The new condition may persist (Fig. 6B).

The most useful suggestion to explain the reorientation is that an unusual stretch rate brings on the cytoskeletal response. When a cell is in a location where it is rapidly stretched by the growth of surrounding organs it often divides, and then reinforces, in the direction of stretch. Presumably, the cell can transduce the fact that it is expanding faster than normal (*e.g.* organelles spread out faster than they can be produced) into this response. A correlation between rapid directional stretch and subsequent reinforcement in that direction has been seen in several cases (9, 11). The correlation with division plane has been noted by others (3). More direct study of this correlation can now be made with sequential scanning electron microscopy (22).

Cells may continue to reinforce in the direction of imposed stretch, as long as the latter persists (B in Fig. 6). Once the effect of exogenous stretch is alleviated, the cell grows normal to its reinforcement, the typical relationship. Thus, a given cell and its progeny can exist in two states: an active one (A) where extension occurs normal to reinforcement and a passive one (B) where the cell is being extended in the direction of its reinforcement. Groups of cells in these alternative states appear to be so positioned in space and so phased in time that a two-dimensional developmental engine results.

New Hoop Formation. Most appendages arise as ridges at the periphery of the dome, "leaf buttresses." The attainment of hoop reinforcement on a ridge involves a 90° shift in cellulose alignment, the new direction being radial to the center of the dome. It is postulated that cells subject to high directional stress, in this case across the crest of the ridge, divide and subsequently reinforce parallel to that stretch. Once the 90° contrast in alignment has been established inside a field of reinforcement lines concentric with the main axis, the components of a new local region of hoop reinforcement are present (Fig. 7A). Cells at the sharp corners of the pattern divide so that alignment in their progeny smooths out the discontinuity. The resulting pattern is one of concentric ellipses. Now the organ can elongate coherently because all cells extend normally to their reinforcement. The new organ is then considered established, biophysically. This sequence appears to apply to many appendages, vegetative and floral, except stamens. The stamen sequence will be dealt with later, under floral development.

Having addressed the principles for producing laterals, we now deal with the progression of large-scale patterns of appendages during vegetative and floral development of a given shoot. An idealized sequence for a crassulacean shoot modeled mainly on *Kalanchoe* will be presented. Information on *Kalanchoe* is from the work of Dr. Amy Nelson. Some details are taken from *Echeveria* where the birefringence patterns are clearer (8). The scheme is a provisional model. The sequence here is especially simple in that the patterns are highly orthogonal, involving angles of either 90° or 45° between appendages.

7 Reinforcement Field Behavior - Top View

A. In Isolation B. In Concert

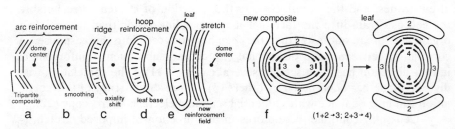

FIG. 7. (A) The elementary unit of decussate apex development, in isolation. A tripartite reinforcement region merges to form a single curved reinforcement field. This arches up into a leaf primordium. An axiality shift occurs on the crest to give hoop-reinforcement to the leaf. The leaf grows in girth to bring about, via cytoskeletal responses to stretch, a gently curved reinforcement field on the dome interior to the leaf. (B) The operation of four such units, properly phased in space and time, generates the vegetative cycle. The central part of an old field (leaf 1) combines with two flank regions of younger fields (leaf 2), to form a new tri-partite reinforcement region (leaf 3). The outer-most large hoop becomes stem. These phenomena account for the repeated production of a new stem segment, two opposite leaves, and major features of a remodeled dome.

A VEGETATIVE-TO-FLORAL SEQUENCE

Vegetative Growth

In a decussate apex, such as that of *Kalanchoe*, a pattern-producing cycle is readily envisioned as in Figure 7. Detailed data are available for *Vinca* (15). It has been pointed out earlier that bulges occur at regions which are distant from the dome center and which have high curvature in the surface plane. The bulge becomes a primordium (Fig. 7A). This could occur in cyclic fashion to explain the decussate pattern. An important element of the proposed mechanism is that a leaf primordium expands in girth. This has two consequences: *i*) the dome is deformed parallel to the leaf base and, hence, is given a new major axis; and *ii*) the dome cells adjacent to the leaf base are stretched, especially near the flanks of the leaf base, so that reinforcement on the dome is parallel to the leaf base (Fig. 7A). This creates a <u>reinforcement field</u> on the dome interior to the leaf. Its lines of curvature are <u>gentle</u>. Bulge initiation requires <u>sharp</u> curvature. This is provided when several such elements act in concert, properly phased (Fig. 7B).

After two opposite leaves have stretched the dome, there are two new sites relatively far from the center. Each site has sharp reinforcement curvature through a combination of parts of three older reinforcement fields. As shown in Figure 7B, the center of the new tripartite field for leaf #3 comes from the central part of the field from leaves of generation #1. The two flanks of the new field come from flanks of fields from generation #2. Thus, a single field comes from parts of fields from two generations of older leaves; a new field subsequently contributes to two later generations of leaves.

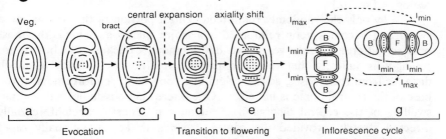

FIG. 8. During the transition to flowering, the vegetative cycle persists, but apparently with less influence from the appendages. Reinforcement is less ordered at the dome center (a to c). At the moment of transition, the center of the dome swells and produces a strong circular reinforcement field around it (d). Parts of this field, plus fields interior to the bracts, provide areas of parallel reinforcement which undergo an axiality shift (e). This produces a maximum inflorescence: bract, I-min, flower primordium, I-min, bract. Each I-min develops into an I-max to generate the cymose inflorescence (dichasium) as seen in Fig. 10.

Because lateral buds appear only later, the vegetative apex has three meristematic humps along a line (leaf, dome, leaf). By the same token, when new creases appear, demarcating the inner side of appendages, they are at 90° to the previous pair of creases (Fig. 2). A major departure from this self-perpetuating vegetative pattern leads to flowers.

The Transition to Flowering

The Shift in Dome Geometry. During the floral transition, two trends modify the above pattern of development. First, leaves get smaller, becoming bract-like (Fig. 8a-c). The appendages appear to have less influence on the dome. Their reinforcement fields no longer extend close to the center of the dome. Secondly, and perhaps as a consequence of the first trend, reinforcement in the central area becomes more random as the decussate pattern continues (Fig. 8c). Just before the actual transition to flowering, there is a large central area, random in its center, but circumferentially reinforced at its periphery (Fig. 8d). One would expect a single giant dome to arise, filling the area between the bracts. Remarkably, only the central portion swells as a dome; it develops circumferential reinforcement (Fig. 8d). Sizable regions remain between the bracts and this swollen central dome. These regions have reinforcement tangential to both the bract and central dome, hence normal to the long axis of the whole meristem, in top view. These regions swell as transverse ridges and, like leaf primordia, undergo an axiality shift along the crest (Fig. 8e). This gives them hoop reinforcement. These are young inflorescence meristems (I-min). The result is that the established meristem, I-max, now has _five_ hoop-reinforced domes in a row: bract, I-min, central flower primordium, I-min, bract (B, I-min, F, I-min, B) (Fig. 9). With time, each I-min becomes I-max. (Compare Fig. 9 with the large central oval region of Fig. 10.)

Why only the randomly reinforced central portion of the original area swells and becomes hoop reinforced is a major question. This is the most obvious difference from vegetative development. Because the bracts are small, it is unlikely that they can provide reinforcement fields at much distance from their bases, so the origin of strong hoop reinforcement at the base of the central dome is not clear. One interesting possibility is that rapid swelling of the central disorganized region can generate circumferentially directed stretch, provided the center expands in an exceptionally rapid manner (10). This expansion could provide a self-organizing feature to the margin of the central area. One assumes that the cells there respond to strong directional stretch in the same way as do cells on the crest of primordium ridges to produce new reinforcement. At any event, the localized swelling, with hoop reinforcement forming at its base, is a major biophysical event in the transition to flowering.

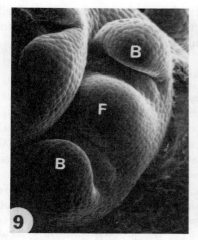
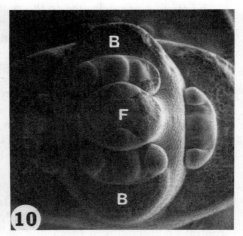

FIG. 9. Structure of flowering apices of *Kalanchoe*. Scanning electron micrograph (SEM) of an inflorescence meristem (I-max). Three major humps lie in a row, diagonally. The end humps are bracts (B); the central mound is a flower primordium (F). Between the flower and the bracts are two young inflorescence meristems (I-min). X 140. SEM courtesy of Drs. Leslie Sunell and Amy Nelson.

FIG. 10. Structure of flowering apices of *Kalanchoe*. An SEM. The large vertical oval is a developed inflorescence apex. The central flower (F) has developed four sepals; two are removed. The large bracts (B) at the ends of the oval have been removed. Two new inflorescence meristems, I-max as in Fig. 9, lie between the flower primordium and the bract sites. X 50. SEM courtesy of Drs. Leslie Sunell and Amy Nelson.

Biophysics of the Transition - Antecedent Causes? An inward shift in the location of the dominant expansion activity appears to characterize the transition to flowering. In vegetative development, the appendages are large and vigorously growing. The entire dome surface can consist of reinforcement fields associated with the leaves of the last two plastochrons. The dome surface appears to be largely passive, its reinforcement structure reflecting the activity of the adjacent leaf bases. The apical surface is small and relatively flat, like a park amid skyscrapers.

The special biophysical aspects of the transition are (*i*) a shift of prominent expansion from the periphery to the central area; (*ii*) randomization of reinforcement near the center, probably a natural consequence of isotropic expansion there; and (*iii*) local "eruption" at the very center to give, typically, a pre-flower mound. The later two features could be brought on by the first. The potential for (*ii*) and (*iii*) may always be present, but can occur only after (*i*) has taken place. It follows that a common consequence of the myriad successful environmental regimes that bring on flowering (1, 2) could be a shift of growth-promoting nutrients or hormones from being delivered to the periphery to being delivered to the center of the apical meristem. A small-scale "nutrient diversion" may be a link between the biochemistry of induction and the biophysical events of the transition.

The Inflorescence Cycle

This cycle has clear-cut similarities and differences relative to the vegetative cycle. The young vegetative meristem produces a stem segment, a leaf pair, plus "itself" at 90°. The young elliptical inflorescence meristem (I-min) produces a stalk, a pair of bracts, a central flower, plus "itself" at 90° twice! (A preoccupation with "itself" has long been noted in the monocot *Narcissus*.)

The new I-min is an elongate ellipse, hoop reinforced. Two bracts form at the ends, presumably in the same fashion as do two leaves in decussate phyllotaxis. The central region, initially with random reinforcement, forms a central hoop-reinforced bulge, just as in the transition to flowering. Two humps form between the central bulge and the bracts, giving five humps in a row. Thus, I-min becomes: B, I-min, F, I-min, B, repeatedly (Fig. 8, f and g; Fig. 9). It is noteworthy that the unusual biophysical phenomena of the transition to flowering, formation of a central hoop-reinforced mound, is repeated twice in every subsequent I cycle. It is also repeated each time a vegetative lateral bud, with two opposite leaves, becomes floral.

The fact that a single I-min meristem produces two new I-min meristems explains how the dichasium inflorescence branches in the manner of a decussate shoot with released lateral buds. For a while, I meristem activity is as cyclic as the vegetative cycle: the same configuration is repeatedly obtained. Ultimately, the new I meristems fail to develop and the inflorescence is complete.

Flower Formation

The biophysical starting point for flower formation is the central mound in the middle of the inflorescence meristem. The mound has hoop reinforcement at the periphery, but has lost cell pattern and reinforcement direction in the center (*Vinca*, 6; *Kalanchoe*, Nelson and Green, in preparation). The dome is not entirely free of the influence of previous structure, however, because the symmetry of sepal formation is related to the previous vegetative phyllotaxis. In *Kalanchoe*, the first pair of sepals have their laminae at 90° to those of the previous bracts, to start a typical flower with 4-fold symmetry. This "feed forward" influence from vegetative phyllotaxis could be in the form of residual reinforcement fields from older appendages. SEM images of *Kalanchoe* weigh against shape being important because the dome appears relatively round (see Fig. 9).

Sepal Formation and the Transition to Four-Fold Symmetry. In *Kalanchoe*, as noted above, the first sepals appear as a pair, normal to the previous bracts (Fig. 11, b and c). Their presence gives the dome a somewhat square outline, as viewed from above. The second sepal pair arises at 90° to the first, on the "edges" of the now more square dome (Fig. 11, c and d). In both cases, an axiality shift gives the sepal its hoop reinforcement. Each sepal generates a reinforcement field interior to itself, on the dome.

Two events occur during sepal formation which shift the angle between appendages at a node from 180° (vegetative) to 90° (floral). First, no internode is formed between the first two pairs of sepals. This could be because the "edge" of the dome giving rise to the second sepal pair is at the same height as the base of the first sepal pair. Second, there is rapid equalization of sepal size of the two generations. Sepals grow extremely rapidly at first, then slow down. This tends to reduce differences in sepal size, but some other factor must serve to fully eliminate it; the difference is gone when the sepals are still growing (Fig. 11e). Perhaps large sepals stimulate the growth of adjacent smaller ones. After this important equalization, the dome is fully "square" in top view (Fig. 10, center). This means that the longest radii on the surface will now lie at 90° intervals rather than the 180° found on the vegetative ellipse.

After sepal formation there is an apparent reversion to mechanisms closely related to the vegetative stage. The dome shape has changed, but the "rule" of placing new appendages at regions of high reinforcement curvature, and along radii of maximal length still holds.

Petals. The four corners of the pre-petal reinforcement band, interior to sepals, rise up as ridges (Fig. 11, e and f). After an axiality shift to acquire hoop reinforcement, each petal primordium forms a reinforcement field, tangential to its base, on the apical dome.

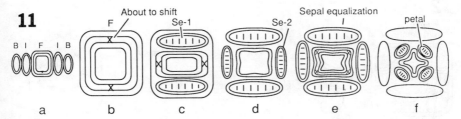

FIG. 11. Development of the flower. Sepals and petals. The rounded dome has an axiality shift occur on the edges normal to the bracts. These areas become sepals (Se-1). After these enlarge, a second pair (Se-2) is produced at the same level on the dome (d). These equalize in size to establish 4-fold symmetry (e). Axiality shifts at the corners of the now square dome establish four petals (f). The cellulose reinforcement pattern for f is taken from polarized light images as in Fig. 3.

Stamens. At the four sites between petal primordia, a whorl of stamens arises. The attainment of hoop reinforcement for these organs appears to be significantly different from that for other floral organs. Rather than having the new hoop reinforcement being the result of an axial shift occurring on a primordium ridge, rudimentary hoop reinforcement is present before bulging. The required circumferential component appears to come from the reinforcement field interior to an organ two cycles old (*e.g.* sepal). This component occupies the center of the new stamen pattern (Fig. 12b). Completion of the new hoop reinforcement is achieved by employing the radially oriented reinforcement lines from the organs on either side (Fig. 12, b-d). The result is that the tip or crest of the stamen has reinforcement lines running circumferential to the dome center. Other appendages, both vegetative and floral, typically have the reinforcement lines on their organ crest running radially. This stamen reinforcement arrangement, the inverse of the typical one, correlates with the stamen having unusual bilateral symmetry at the tip. In the anther, a major plane of symmetry, the partition between locule pairs, is radial to the dome center. The reinforcement in this region appears to be descended from a field formed two cycles earlier (*e.g.* sepals). The base of the stamen, the filament, has radial symmetry. Thus, some unique features of stamen symmetry correlate with an unusual biophysical mechanism of attaining hoop reinforcement. The nature of the organs giving rise to the pertinent reinforcement fields for stamens does not seem to matter. For the first whorl of stamens, the central field comes from a sepal; for the second whorl, it comes from a petal (Fig. 13). Stamens generate reinforcement fields with alignment tangential to their base. After merging, these fields are important for carpels.

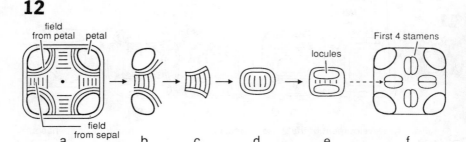

FIG. 12. Development of the flower. The first stamens. New reinforcement fields are tangential to the petals. Old central fields from the sepals persist between petals. Parts of three fields combine to provide primitive hoop reinforcement for a stamen (b, c). The reinforcement on the crest of the stamen is circumferential to the floral dome (different from other organs). A plane of symmetry of the stamen, separating locule pairs, is radial.

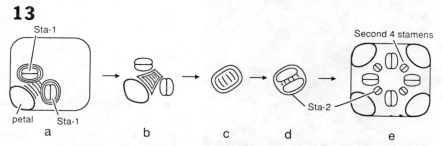

FIG. 13. Development of the flower. The second stamens. The process in Figure 12 repeats, but with the central field coming from a petal, the lateral fields from first stamens.

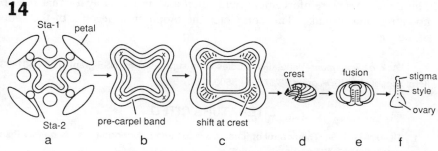

FIG. 14. Development of the flower. Carpels. The reinforcement fields from the two sets of stamens merge to form a four-leaf clover-shaped pre-carpel band. Axiality shifts occur at the curved peripheral regions. These crests arch up and ultimately fuse shut. The hoop-reinforced hollow carpel arises and matures (d,e,f).

Carpels. In the crassulacean flowers studied so far, the carpels arise from a sinusoidal reinforcement band formed by a merging of the reinforcement fields just interior to the recent whorls of stamens. The band has the outline of a four-leaf clover. This reflects the pattern of the stamens (Fig. 14a). As before, the most peripheral parts of the formative region are the most active. These "tips" of the band arch up. The initial arching of the band is seen in Figure 15. These regions undergo an axiality shift to provide the radial component of reinforcement (Fig. 14c). This appears to take place ultimately over almost the entire band. As the tips of the band arch up, the original crest of the ridge, the area of the axiality shift, becomes the rim of the now-closing carpel opening (Fig. 14, d and e). The opposing edges of this opening fuse to make the hollow carpel. The tip of this structure, presumably including the region of the axiality shift, becomes the style and stigma (Fig. 14f). The small area interior to the carpels is not stretched by carpel growth.

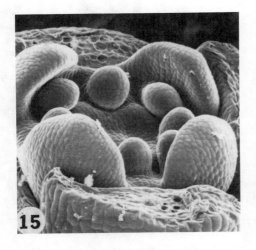

FIG. 15. Structure of flowering apices of *Kalanchoe*. An SEM of a flower which is starting to form carpels interior to the eight stamens. Sepals have been removed. X 125. SEM courtesy of Drs. Leslie Sunell and Amy Nelson.

There are no new reinforcement fields on it and, presumably for that reason, no subsequent organs. This completes the biophysical account of the shoot sequence.

Acknowledgement--Vegetative studies were supported by NSF grant DCB-8801493; floral aspects by USDA grant 88 37261 - 3500.

LITERATURE CITED

1. BERNIER G 1988 The control of floral evocation and morphogenesis. Annu Rev Plant Physiol 39: 175-219
2. BERNIER G, J-M KINET, RM SACHS 1981 The Physiology of Flowering, Vol II. CRC Press, Boca Raton, Florida
3. FRENCH JC, DJ PAOLLILO JR 1975 The effect of the calyptra on the plane of guard cell mother cell division in *Funaria* and *Physcomitrium* capsules. Ann Bot 39: 233-236
4. GOLDBERG RB 1988 Plants' novel developmental processes. Science 240: 1460-1467
5. GREEN PB 1984 Shifts in plant cell axiality: division direction influences cellulose orientation in the succulent *Graptopetalum*. Dev Biol 103: 18-27
6. GREEN PB 1985 Surface of the shoot apex: a reinforcement field theory for phyllotaxis. J Cell Sci Suppl 2, pp 181-201
7. GREEN PB 1987 Inheritance of pattern: analysis from phenotype to gene. Am Zool 27: 657-673
8. GREEN PB 1988 A theory for inflorescence development and flower formation based on morphological and biophysical analysis in *Echeveria*. Planta 175: 153-169
9. GREEN PB, KE BROOKS 1978 Stem formation from a succulent leaf: its bearing on theories of axiation. Am J Bot 65: 13-26

10. GREEN PB, A KING 1966 A mechanism for the origin of specifically oriented textures in development with special reference to *Nitella* wall structure. Aust J Biol Sci 19: 421-437
11. GREEN PB, JM LANG 1981 Toward a biophysical theory of organogenesis: birefringence observations on regenerating leaves in the succulent *Graptopetalum paraguayense*. Planta 149: 181-195
12. GUNNING BES, AH HARDHAM 1982 Microtubules. Annu Rev Plant Physiol 33: 651-698
13. HARRISON LG, M KOLAR 1988 Coupling between reaction-diffusion prepattern and expressed morphogenesis, applied to desmids and dasyclads. J Theor Biol 130: 493-515
14. HOLDER N 1979 Positional information and pattern formation in plant morphogenesis and a mechanism for the involvement of plant hormones. J Theor Biol 77: 195-212
15. JESUTHASAN S, PB GREEN 1989 On the mechanism of decussate phyllotaxis: biophysical studies on the tunica of *Vinca major*. Am J Bot (in press)
16. KUTCHERA U 1987 Cooperation between outer and inner tissues in auxin-mediated plant organ growth. *In* DJ Cosgrove, DP Knieval, eds, Physiology of Cell Expansion During Plant Growth. Am Soc Plant Physiologists, pp 215-226
17. LLOYD CW 1982 The Cytoskeleton in Plant Growth and Development. Academic Press, New York
18. MEINHARDT H 1982 Models of Biological Pattern Formation. Academic Press, London
19. OSTER GF, JD MURRAY, AK HARRIS 1983 Mechanical aspects of mesenchymal morphogenesis. J Embryol Exp Morph 78: 83-125
20. SAKAGUCHI S, T HOGETSU, N HARA 1988 Arrangement of cortical microtubules in the shoot apex of *Vinca major* L. Observations by immunofluorescence microscopy. Planta 175: 403-411
21. SAKAGUCHI S, T HOGETSU, N HARA 1988 Arrangement of cortical microtubules at the surface of the shoot apex in *Vinca major* L.: observations by immunofluorescence microscopy. Bot Mag Tokyo (in press)
22. WILLIAMS MH, PB GREEN 1988 Sequential scanning electron microscopy of a growing plant meristem. Protoplasma 147: 77-79

ADDENDUM

The biophysical account given above addresses the issues of organ number, placement, and orientation. The possible link between biophysics and organ identity may also be briefly considered.

The floral state clearly influences the pattern in which hoop reinforced appendages are made at the apex. But what determines the character of the appendages produced? The early reinforcement prepattern for some of the organs appears to be distinctive. For example, the pattern for stamens is nearly circular; that for carpels is a wavy band. The initial pattern goes through several rounds of further modification to yield the physical structure of the final organ.

Initiation of a Specific Organ. It is conceivable that the succession of organ types in a flower could have a strictly biophysical basis. One cycle of organ formation could modify the reinforcement pattern on the dome, and the dome's three-dimensional shape, to generate a distinctive structure prepattern for the next cycle. That this simple picture is not correct is indicated by the appearance of organs out of sequence. Lyndon (3) reports carpels forming on proliferous flowers of *Silene* which lack antecedent petals or stamens. Lu *et al.* (4) report the appearance of many ovules in cultured *Hyacinthus*, with no antecedent ovaries. Further, the proportion of stamens to ovules in culture is greatly modified by the type of cytokinin used in the medium. This indicates that the biochemical state of the tissue can influence organ character. There is presumably biochemical influence on the early biophysical events (*e.g.* whether the hoop reinforced area is to be round or elongate). Similar "unscheduled" (no normal biophysical precursor) origins of distinctive new axes are also seen in embryo culture where adventitious embryonic axes appear repeatedly. Granting that the biochemical background can strongly influence the early reinforcement pattern for an organ, there remains the question of what determines whether the same organ type is repeated many times, as it can be with stamens (2) or a new organ type is produced. A second question is whether a specific early biophysical pattern inevitably produces a specific organ type.

Development of a Specific Organ. It has been suggested that development of an organ involves successive rounds wherein a specific structure (brought on by biophysics) induces specific biochemistry and the biochemistry then influences subsequent biophysics. Such a cycle could differentiate anther from filament. The final biochemical inductions would yield the major site and stage specific gene activations. One might expect that a specific initial biophysical structure could only lead to a given cascade of interactions and hence the corresponding cytodifferentiation for the organ types. This expectation is not met. What can be interpreted as a dramatic shift from one cascade to another is shown by Rasmussen's (unpublished data; see abstract this volume, p. 180) study of the <u>green pistillate</u> mutant of tomato. Here, stamens originate in normal positions by presumably normal biophysical events. After acquiring the typical slight bilobed appearance, the stamens adopt the carpel sequence of development. They fuse laterally, the ring of stamens becoming like a giant carpel surrounding and fused to the normal gynoecium. The fused stamens develop styles and stigmoid tips. Ovules appear in the space between the former stamens and the gynoecium. The stamens, despite never undergoing the biophysical early development of a carpel (forming a curved ridge which fuses in on itself), show striking carpel character, including red pigmentation. Thus, the sequential biophysical aspects of organ development, as well as organ initiation, appear subject to biochemical control. One hormone, gibberellic acid, is known to influence biophysics and, hence morphogenesis through an action on microtubules (1). It may be one of many. The levels of such effectors can clearly influence subsequent biophysics; it is suggested here that the levels themselves are normally in some way the product of antecedent biophysics. In the case of green pistillate, the original stamens are diverted to a carpel cascade presumably by excess of carpel effector or an aberration in the stamens' response system.

The present view, that progressive rounds of biophysical-biochemical interaction may account for organ identity, shares with traditional reaction-diffusion explanations the idea that pattern is generated at successively smaller levels of scale. Our view differs in two ways. First, in the present scheme, the presence or absence of an effector is the critical feature, not the geometry of the effector's concentration profile. Effectors act on complex structure, do not themselves generate it. Secondly, and accordingly, the physical construction process is thought to be involved in all levels of pattern generation, rather than to be a late response to a detailed chemical prepattern.

LITERATURE CITED

1. MITA T, H SHIBAOKA 1984 Gibberellin stabilizes microtubules in onion leaf sheath cells. Protoplasma 119: 100-109
2. HUFFORD LB 1989 The evolution of floral morphological diversity in Eucnide (Loasaceae). The implications of modes and timing of ontogenetic changes on phylogenetic diversification. In P Liens, SC Tucker, PK Endress, eds, Aspects of Floral Development. Gebrüder Borntraeger, Berlin-Stuttgart (in press)
3. LYNDON RF 1979 A modification of flowering and phyllotaxis in Silene. Ann Bot 43: 553-558
4. Lu W, K Enomoto, Y Fukunaga, C Kuo 1988 Regeneration of petals, stamens and ovules in explants from perianth of Hyacinthus orientalis L. Importance of explant age and exogenous hormones. Planta 175: 478-484

GENE EXPRESSION DURING FLORAL INITIATION

D. R. MEEKS-WAGNER, E. S. DENNIS, A. KELLY, S. SHANNON, R. WHITE, J. WAHLEITHNER, A. NEAL, M. LUND, AND W. J. PEACOCK

Institute of Molecular Biology, University of Oregon, Eugene, OR 97405, USA, (D.R.M-W., A.K., S.S., R.W.); *and Division of Plant Industry, CSIRO, GPO Box 1600, Canberra, ACT, 2601, Australia* (E.S.D., J.W., A.N., M.L., W.J.P.)

The extensive physiological and morphological characterization of flowering has provided a wealth of information describing this important developmental process. This information has lead us to view the transition to flowering as a progression of steps in a pathway that leads to the commitment of the shoot meristem(s) to form flowers. Once committed to floral development, in normal circumstances, these meristems produce floral organs in sequential order to give rise to fully developed flowers. It has long been proposed that alterations in gene expression are associated with the transition to flowering (for example, 13), and given the recent advances in molecular biology, we are now able to identify and isolate genes whose activity changes during developmental processes. This paper will summarize our recent studies of changes in gene expression during floral initiation.

Most of the information about the flowering process has been derived from studies with a relatively few number of plant species, the majority of which display some dependence upon a particular photoperiodic regime in order to flower. The advantages of using a photoperiodic-dependent system for investigating flowering are obvious. Of equal interest is the mechanism of floral initiation in day-neutral (DN) plants which flower without requiring a particular photoperiodic treatment; such plants often appear to have an "internal developmental clock" which programs them to flower at a given developmental or physiological age. The various species of *Nicotiana* offer an experimental system possessing both DN and photoperiodic plants in which floral induction and floral meristem determination have been well studied (2, 5, 6, 8). In addition, Tran Thanh Van (10) has developed an *in vitro* organogenesis system with DN *Nicotiana tabacum* that permits the direct formation of floral meristems and flowers on thin strips of tissue taken from the floral branches of fully flowering plants. We have used this system, the tobacco "thin cell layer" (TCL) system, to isolate genes expressed early during *in vitro* floral initiation, and have characterized the transcriptional expression of some of these genes during normal plant development.

EXPERIMENTAL SYSTEM

TCL taken from the floral branches of *Nicotiana tabacum* cv samsun can be induced to follow different programs of organogenesis, depending on the composition of the culture medium and the culture conditions (11). Using culture conditions described previously (12), we were able to induce floral bud (FB) or vegetative shoot (VS) formation within 20 d of culturing the TCL explants on liquid medium. This switch in developmental programs was controlled by the type of cytokinin included in the medium: kinetin stimulated direct flower formation, while zeatin stimulated VS production; approximately 20% of these VS formed FB after producing three to four leaves (about 45 d after the initiation of the cultures). The type of cytokinin was the only variable component of the TCL cultures.

We also made use of *Nicotiana* plants grown from seed in the greenhouse for the characterization of gene expression during normal plant development. *N. tabacum* cv samsun is a DN plant in which developmental age can be estimated by counting the node number (8). *N. sylvestris* is a LD species, and *N. tabacum* cv Maryland Mammoth is a SD variety derived from DN tobacco (1, 5). Experiments with photoperiodic plants were performed by first growing the plants in noninductive conditions until they were competent to respond to the inductive photoperiod, and then transferring plants to an inductive environment until they flowered.

RESULTS AND DISCUSSION

Isolation of Floral TCL cDNA Clones. A cDNA library was made from Poly(A)+RNA isolated from day 7 FB explants in the bacteriophage λ gt10 (4). The unamplified library was screened with cDNA probes made from day 7 FB (FB7) explants and day 7 VS (VS7) explants. This screen resulted in 52 cDNA clones being identified as expressed in FB7 explants, and only weakly expressed in VS7 explants. These 52 clones were shown to represent six different gene families, termed FB7-1, 2, 3, 4, 5, and 6. FB7-1, 2, 3, and 4 were each cloned more than 10 independent times, while FB7-5 was cloned twice and FB7-6 was cloned once.

Initial characterization of these genes was done by dot blot analysis using cDNA probes made from various TCL experiments. These probes included: time-zero TCL explants (*i.e.* explants harvested prior to culture), day 20 floral TCL explants (FB20), and TCL explants taken from the basal portion of the plant stem designated intermediate zone (non-inflorescence) (IMZ) and cultured on floral (*i.e.* kinetin) medium for 7 d (IMZ7). The IMZ7 material was a test for gene expression induced simply by the culture conditions, as explants taken from the lower stem from VS rather than FB when cultured on floral medium (11). This experiment showed that the expression of the FB7 genes we isolated was induced in floral TCL explants, and persisted in these explants at least through day 20 of culture. Only FB7-3 was significantly expressed in IMZ7 explants, suggesting that the induction of the FB7 genes was correlated with the initiation of flowering and not simply the result of kinetin in the culture medium.

Further Characterization of FB7-1, 2 and 5. Genes FB7-1, 2 and 5 were shown to exist in low copy (or single copy) number in the tobacco genome by Southern analysis, and Northern analysis demonstrated that each gene encodes a single major unique RNA; thus, these genes were selected for further characterization of their patterns of transcription during *in vitro* and *in vivo* floral initiation.

FB7-1, 2, and 5 displayed similar patterns of expression during *in vitro* organogenesis. Maximum expression was detected in FB7 explants; very little activity was found in day 13 FB (FB13) explants, and approximately 50% maximum activity was found in day 25-33 FB (FB25-33) explants. Low activity was found in VS7 and day 13 VS (VS13) explants, but high levels of expression (approximately equal to those observed in FB7 explants) were detected in day 25-33 VS (VS25-33) explants. The high level of expression in VS25-33 explants is interesting in light of the fact that VS25-33 explants are at a stage which precedes FB formation by about 14 d (*i.e.* it is somewhat equivalent to day 7 of FB culture).

The transcriptional expression of FB7-1, 2, and 5 was examined during normal plant development by *in situ* hybridization and Northern analysis. *In situ* hybridization indicated that FB7-1 and FB7-2 were transcriptionally expressed at significant levels in the subapical pith cells of *N. tabacum* cv samsun plants possessing pre-floral meristems or immature inflorescences; these genes were expressed only at much lower levels in the apices of vegetative plants. Northern analysis revealed that transcripts of FB7-1, 2, and 5 were most abundant in the roots of mature *N. tabacum* cv samsun plants, and that lower levels of these transcripts were detectable in older leaves and internode tissue. FB7-1 transcript was detectable in floral branch and unopened flower Poly(A)+RNA populations; FB7-2 was detectable at a high level in floral branch RNA, but not in unopen flower RNA, and FB7-5 was not detected at significant levels in either of these samples.

Temporal Pattern of FB7-1, 2, and 5 Gene Expression in Roots. The temporal pattern of FB7-1, 2, and 5 gene expression in the roots of DN and photoperiodic *Nicotiana* was further investigated using Northern analysis. During the growth of DN *N. tabacum* cv samsun, the steady-state levels of RNA for FB7-1 and 2 (and to a lesser degree FB7-5) increase with age until the plant reaches a developmental stage when the shoot meristem becomes florally determined. [The commitment of the shoot meristem to floral development was tested under our growth conditions by removing the upper internodes and apex of plants and observing the developmental fate of the meristem in isolation from the rest of the plant (see 8).] The transcriptional activity of these genes then decreased to approximately 20% of the maximum level as the development of the immature inflorescence proceeded. Expression of these genes may oscillate around this lower level throughout the remaining developmental stages, but we have not looked at enough late stages to draw any conclusions at this time.

Observing this temporal pattern of gene expression during the life cycle of DN *Nicotiana tabacum* cv samsun suggested that more information about the developmental control of FB7 gene activity could be obtained by investigating the levels of FB7 transcripts in photoperiodic *Nicotiana*. We have examined the expression of FB7-2 and FB7-5 in the roots of *N. sylvestris* (a LD plant) and *N. tabacum* cv Maryland Mammoth (a SD plant) by comparing adult vegetative plants (*i.e.* plants capable of floral initiation if grown in the proper photoperiod) and plants possessing an immature inflorescence. FB7-2 is expressed at much higher levels in the roots of adult vegetative *N. sylvestris* plants than in the roots of *N. sylvestris* plants possessing an immature inflorescence; FB7-5 is expressed at equal levels at both of these developmental stages in *N. sylvestris*. Experiments with *N. tabacum* cv Maryland Mammoth plants have demonstrated that there is little, if any, differential expression of FB7-2 or FB7-5 between the adult vegetative stage and the immature inflorescence stage. The difference in the pattern of FB7-2 expression observed in these two *Nicotiana* species may be related to a difference in growth events associated with the transition of adult vegetative plants to the floral state: *N. sylvestris* occurs as a compact rosette plant in the adult vegetative state and rapidly elongates with floral induction, while adult vegetative *N. tabacum* cv Maryland Mammoth plants possess elongate stems which do not substantially increase following the induction of flowering.

Identification of FB7-1, 2 and 5. FB7-1, 2, and 5 have recently been identified by DNA:DNA homology and DNA sequencing as being homologous to some of the pathogenesis related (PR) genes of *Nicotiana*. Southern analysis has demonstrated that FB7-1 is homologous to chitinase and that FB7-5 is homologous to β-glucanase. FB7-2 shares DNA sequence homology to PR-S.

CONCLUSIONS

The tobacco TCL system has provided an efficient method for collecting vegetative and floral tissue at equivalent developmental stages. The ability to control TCL organogenesis with common plant growth substances may permit the mode of action of these substances to be further defined. However, one major limitation of the system is that much like altering the photoperiodic regime of plants to induce flowering, with the TCL system at least one component of the culture conditions must be altered to produce different developmental programs for comparison. These alterations need to be taken into consideration when evaluating the results of such comparative studies.

We have demonstrated that the FB7 transcripts are induced to high levels in floral TCL explants during the first 7 d of culture, while they are present in only low amounts in VS TCL explants. With the possible exception of FB7-3, the FB7 genes are not highly expressed in TCL explants derived from lower regions of the tobacco stem when these explants are cultured on floral medium; these IMZ explants form VS instead of FB on the kinetin medium and, thus, this experiment suggests that the increase in FB7 transcripts observed in floral explants is not due simply to the presence of kinetin in the floral culture medium.

The observations on FB7 gene expression during normal plant development suggest that physiological changes during plant growth may trigger the coordinated expression of some of these genes. Given the identity of FB7-1, 2, and 5 (*i.e.* showing homology to PR genes), it is unlikely that these genes are directly involved in the transition to flowering. However, the expression of such genes has been shown to be influenced by developmental changes in healthy plants (7, 9), as well as by nonpathogenic stress [such as that induced by ethylene, (3)] and by altering the endogenous levels of cytokinins in plants (9). The creation of conditions, both *in vitro* and *in vivo*, which influence the flowering process may also effect the expression of the FB7 genes. Future experiments will test this hypothesis in order to determine if the FB7 genes can serve as probes for early events in the transition from vegetative to floral development.

LITERATURE CITED

1. ALLARD HA 1919 Gigantism in *Nicotiana tabacum* and its alternative inheritance. Am Nat 53: 218-233
2. BERNIER G 1988 The control of floral evocation and morphogenesis. Annu Rev Plant Physiol Plant Mol Biol 39: 175-219
3. BROGLOE KE, JJ GAYNOR, RM BROGLIE 1986 Ethylene-regulated gene expression: Molecular cloning of the genes encoding an endochitinase from *Phaseolus vulgaris*. Proc Natl Acad Sci USA 83: 6820-6824
4. HUYNH TV, RA YOUNG, RW DAVIS 1985 Constructing and screening cDNA libraries in lambda gt10 and lambda gt11. *In* DM Glover, ed, DNA Cloning: A Practical Approach, Vol 1. IRL Press, Oxford, pp 49-78
5. LANG A 1965 Physiology of flower initiation. *In* W Ruhland, ed, Encyclopedia of Plant Physiology, Vol 15 (Part 1). Springer-Verlag, Berlin, pp 1380-1536
6. LANG A, MKH CHAILAKHYAN, IA FROLOVA 1977 Promotion and inhibition of flower formation in a dayneutral plant in grafts with a short day plant and a long day plant. Proc Natl Acad Sci USA 74: 2412-2416
7. LOTAN T, N ORI, R FLUHR 1988 Pathogenesis related proteins are developmentally regulated in tobacco flowers. Paper (#235) presented at the 2nd International Congress of Plant Molecular Biology, Jerusalem, Israel, November 13-18

8. McDaniel CN, Sr Singer, KA Dennin, JS Gebhardt 1985 Floral determination: timing, stability, and root influence. *In* M Freeling, ed, Plant Genetics. Alan R Liss, New York, pp 73-87
9. Memelink J, JHC Hoge, RA Schilperoort 1987 Cytokinin stress changes the developmental regulation of several defense-related genes in tobacco. EMBO J 6: 3579-3583
10. Tran Thanh Van M 1973 Direct flower neoformation from superficial tissue of small explants of *Nicotiana tabacum* L. Planta 115: 87-92
11. Tran Thanh Van M, N Thi Dien, A Chlyah 1974 Regulation of organogenesis in small explants of superficial tissue of *Nicotiana tabacum* L. Planta 119: 149-159
12. Tran Thanh Van M, P Toubart, A Cousson, AG Darvill, DJ Gollin, P Celf, P Albersheim 1985 Manipulation of the morphogenetic pathways of tobacco explants by oligosaccharins. Nature 314: 615-617
13. Wardlaw CW 1957 The floral meristem as a reaction system. Proc R Soc Edinb 66: 394-408

KINEMATIC ANALYSIS OF LILY FLOWER ORGANS

E. M. LORD AND K. S. GOULD

Department of Botany and Plant Sciences, University of California, Riverside, CA 92521, USA (E.M.L.); *Department of Botany, University of Auckland, Private Bag, Auckland, New Zealand* (K.S.G.)

Plant development studies have focused mainly on the vegetative organs, the root and leaf, with less attention paid to the more complex system of the flower. The flower is determinate in its growth, producing a serial progression of different parts from embryonic tissue, each showing discrete stages of differentiation, making it an ideal system for studying morphogenesis and correlative growth of plant organs. The general assumption to date is that the determinate organs of the flower are homologous to leaves, showing similar patterns of growth (1, 9-11). The work on initiation patterns in flower organs substantiates this claim (3, 4, 16). But, there are few studies that compare flower organ growth after initiation to patterns seen in leaf development.

The bulk of floral morphological studies have been phylogenetic or systematic in their approach, focusing mainly on organ initiation patterns, and few have applied the more quantitative, descriptive methods to the analysis of organ growth for which we have many examples in the literature on the leaf, shoot, and root.

There has been a recent surge of interest in the molecular genetics and biochemistry of flower organ initiation and development. This approach may ultimately resolve the old question about when, during the chain of events leading to evocation and flower organ initiation, new genes come into action and what the specific flower organ developmental genes are. We do not, as yet, have an adequate base of knowledge in the area of floral organogenesis on which to use these sophisticated tools and the information that they will yield about molecular machinery in the flower. This is less true for the better characterized processes of gametogenesis, pollination, and embryo development, which have already attracted intensive efforts by molecular biologists (6, 12).

In order to obtain an adequate picture of morphogenesis in plant organs, we must study them with respect to both space and time. For every stage in development, we need to understand how surface growth features and internal differentiation patterns are contributing to the three-dimensional morphology of the organ as a whole. A detailed picture of the temporal changes in shape and size of floral organs during ontogeny is necessary to reveal the processes by which a collective unit as complex as the flower is achieved.

Surface marking experiments provide a picture of both material (*i.e.* cellular) and spatial aspects of organ growth (20). One applies marks to an organ and observes their displacement during subsequent growth to gain an appreciation for the patterns of organ deformation during ontogeny. Time-lapse photographs of the marked growing organs provide the raw data to calculate local relative growth rates for marked segments. Additionally, displacement velocities from a fixed point in the organ can be calculated to give the relative elemental growth rate. Using such particle distribution studies, both one-dimensional growth along the root axis and two-dimensional growth of a leaf lamina have been described (8). Such kinematic analyses of growth approach the study of motion as a phenomenon and do not pertain to underlying forces (20). If growth is centered in the same region over time, then growth is said to be steady, as occurs in the indeterminate shoot and root. In determinate organs, such as the leaf, growth centers shift during ontogeny and so growth is said to be nonsteady (20).

Analyses of floral organ growth are rare, consisting mainly of Ritterbusch's work on the complex flower of *Calceolaria* where the evolving shape of the petals was described by trajectories from recognizable landmarks (19). In our studies of flower organ growth in lily, we have applied the techniques of kinematic analysis in combination with conventional studies of histology and morphology to provide a picture of flower organ growth from inception to maturity (14, 15).

Lilium longiflorum, easter lily, has a perianth that is not differentiated into calyx and corolla, but rather is composed of six tepals in two whorls (Fig. 1). This is typical of monocotyledons, as is the trimerous nature of the flower. The stamens are six in number, and the ovary is comprised of three carpels fused to form one structure, the gynoecium.

Tepals. The tepals are produced in two whorls alternate with each other. Each inner tepal has a prominent abaxial midrib which protrudes between two adjacent outer tepals. We placed marks along this midrib which acts as a suture, or region of joining, for the outer whorl of tepals (Fig. 2a).

Time-lapse photographs of marked tepals allowed us to visualize the longitudinal patterns of surface expansion from early bud (3.7 mm) to maturity (156.6 mm) just prior to anthesis. There are three phases of growth of lily tepals (Table I). The youngest tepals show both spatial and temporal

FIG. 1. Flower of *Lilium longiflorum* Thumb. T = tepal; S = stigma; A = anther. X 1.

variation in growth rate (phase I) that continues until a 10 mm tepal length (Fig. 2, A-C). Growth during phase I is not diffuse, but rather shows peaks and troughs which occur randomly along the tepal. Phase II growth (tepal length 10-90 mm) is steady and predominantly basal (Fig. 3, A-C). Phase III (tepal length 90 mm to anthesis) shows a shift from basal to apical growth, culminating in anthesis (Fig. 4).

In phase I, mitotic activity occurs evenly throughout the tepal. In phase II, a bimodal distribution of mitotic activity appears with peaks at the base and just above the midpoint of the tepal. Cell division becomes restricted to the base and later ceases when tepals are 40 mm in length, approximately one-third their mature length. It is evident from the marking experiments, that basal growth (phase II) begins before mitoses are restricted to the base and phase II outlasts any cell division in the tepal. For this reason, it is not strictly appropriate to attribute the predominant basal growth seen here in lily to a basal "meristem" *per se*. The term "basal meristem" implies that cell division has become restricted to the base and, hence, is responsible for basal growth. Our data suggest that the growth pattern itself may modify the spatial distribution of mitosis. Of the three phases of growth, only phase II is steady; the other two show spatial variation in growth rate over time. Of the

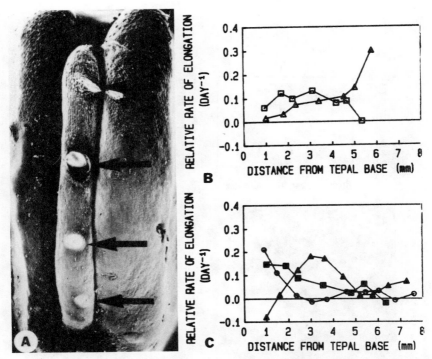

FIG. 2. A, SEM of lily flower bud in phase I (6 mm). The abaxial midrib of an inner tepal is marked with a file of dots of carbon (three tepals are visible). X 6.3. Buds were photographed at time of marking (with reference scale) and then at 24-h intervals over 3-30 d. B, C, Local relative growth rate (LRGR) profiles of a tepal over 6 consecutive days during phase I. Initial length (mm): □ 6.3, day 1-2; △ 6.8, day 2-3; ■ 7.6, day 3-4; ▲ 8.2, day 4-5; ○ 8.7, day 5-6. Fig. 2B,C reproduced from Gould and Lord (1989) with permission from Springer-Verlag.

Table I. *Summary of marking experiments on tepals of Lilium longiflorum*

Data from 217 plots of relative growth rate. The number of tepals from which each set of plots was derived is shown in parentheses.

		Number of Plots				
	Length		Position of Growth Maximum (% Distance from Base)			
Phase	(mm)	Total	0-25	26-50	51-75	76-100
I	3.7-9.9	46 (11)	12	11	16	7
II	10-19.9	32 (8)	22	5	3	2
	20-29.9	49 (15)	37	7	0	5
	30-59.9	36 (10)	32	3	0	1
	60-89.9	20 (8)	15	4	0	1
III	90-119.9	14 (8)	6	3	2	3
	120-150.0	20 (9)	1	3	7	9

Reproduced from Gould and Lord (1989) with permission from Springer-Verlag.

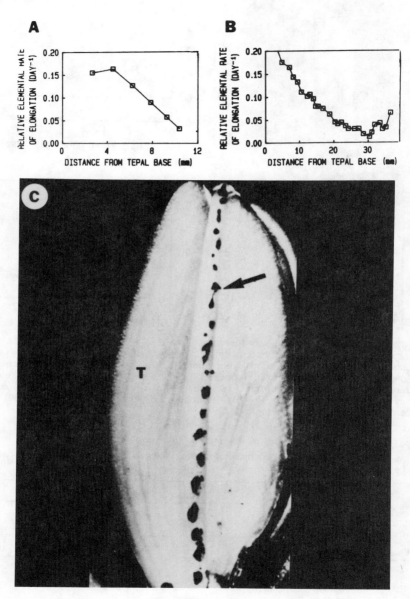

Fig. 3. A, B, Relative elemental rate (RER) profiles for lily tepal during early and late phase II. Initial length (mm): (A) 11.2, day 3; (B) 36.7, day 15. C, Phase II lily tepal (30 mm) marked along the abaxial midrib of an inner tepal which acts as a suture for the two adjacent outer tepals. X 4. Figure 3 A,B reproduced from Gould and Lord (1989) with permission from Springer-Verlag.

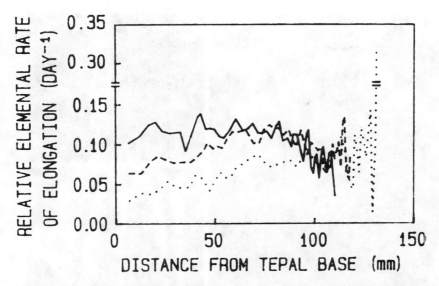

FIG. 4. Relative elemental rate profiles of one tepal over 3 consecutive days during phase III. Length (mm): (a) 111.1 ———; (b) 119.0 -----; (c) 132.2 ..... (anthesis). Figure reproduced from Gould and Lord (1989) with permission from Springer-Verlag.

approximately 45 d from tepal initiation to maturity, 26% of the time is spent in phase I, 53% in phase II, and 21% in phase III (W. Crone, unpublished data).

Stamens. Initiation of the six stamens occurs after that of the tepals (Fig. 5A). A stamen is composed of an anther and a filament (Fig. 5B) at maturity, but only the anther is visible early on (Fig. 5C). The filament becomes intercalated between the point of attachment of the stamen on the receptacle and the anther connective at a later stage in development (Fig. 5D). Locules are initiated in the anther when it is less than 1 mm long (Fig. 5A) and by 2 mm, initiation of the wall layers and sporogenous tissue has occurred. Most of the growth after this is due to the longitudinal extension of the anther until it is 20 to 25 mm long (Fig. 5, B and D).

Since elongation of the tepals is exponential in lily for a considerable period of time (~1 month) (Fig. 6), the bud length can be used as a developmental index by which anther growth can be measured (7). The allometric plot of anther length by tepal length shows two phases of anther growth, the first having a slope (or allometric constant, k) of 1.14, and the second, later phase, a slope of 0.25 (Fig. 6B). Using the relative rate of tepal elongation during anther growth, $r = 0.103 \cdot d^{-1}$, one can calculate relative rates of anther elongation, $kr = 0.118 \cdot d^{-1}$, for the early phase of growth up to 20 mm and for the later decline phase, $kr = 0.027 \cdot d^{-1}$. With these calculated measures of *in situ* anther growth, we could monitor the effects of dissection for marking purposes on normal growth. The growth profiles resulting from 24-h marking experiments (longer intervals were not possible

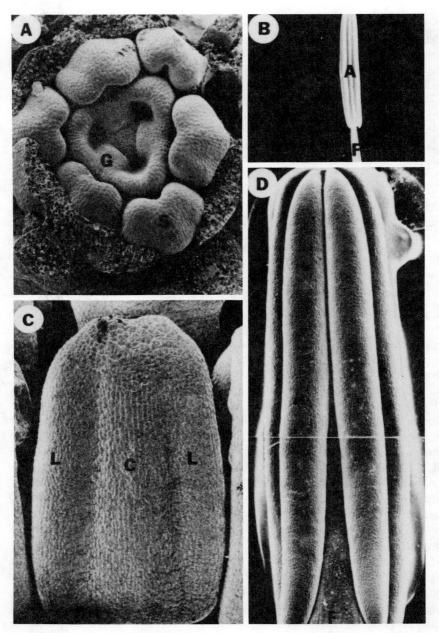

Fig. 5. Stamen development in lily. A, SEM of flower primordium with tepals removed. T_i = inner tepal; T_o = outer tepal; S = stamen; G = gynoecium. X 46. B, Mature stamen of lily. A, anther; F, filament (only upper segment shown; total length = 100 mm). X 1.28. C, Abaxial view of 1.4 mm anther where carbon marks would be placed equidistantly in a file after dissecting a window in the tepals. X 59. D, Adaxial view of 7 mm anther. Note filament initiation. X 19.2.

due to damage from tepal dissection) show that growth is not continuous along the anther, but rather occurs in discrete areas which are shifting constantly in time (Table II; Fig. 7, A-D). In the smaller anthers (Figs. 5C and 7A), there was a single growth peak, and in the larger ones, two or three peaks and troughs could occur (Figs. 5D and 7D). Both the height and width of the peaks varied and the patterns could be different among the six anthers of the same flower. We postulated that the peaks represent regions of predominantly cell expansion and the troughs of predominantly cell division. Getting an anther to grow at the *in situ* rate for more than 24 h after tepal dissection was difficult, but for the few such successful experiments in which

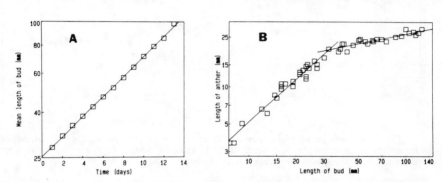

FIG. 6. *In situ* growth of lily floral buds. A, Time course of elongation of tepals. B, Logarithmic plot of anther length against tepal length. Figure reproduced from Gould and Lord (1988) with permission from Springer-Verlag.

Table II. *Summary of data from marking experiment on anthers of Lilium longiflorum*[a]

Anther Length	Total	Trend Not Obvious	Number of Peaks Per Anther			Percent Distance from Base at Which Tallest Peak Occurred			
			1	2	3	0-25	26-50	51-75	76-100
(mm)									
0-2.9	9	0	7	2	0	1	1	4	3
3-5.9	13	0	9	3	1	2	2	6	3
6-8.9	17	2	8	5	2	1	4	3	7
9-11.9	12	2	4	6	0	4	1	3	2
12-14.9	11	2	4	3	2	5	2	2	0
15-17.9	2	0	1	1	0	1	1	0	0
Total	64	6	33	20	5	14	11	18	15

[a] Arrows denote overall basipetal movement of the peak. Reproduced from Gould and Lord (1988) with permission from Springer-Verlag.

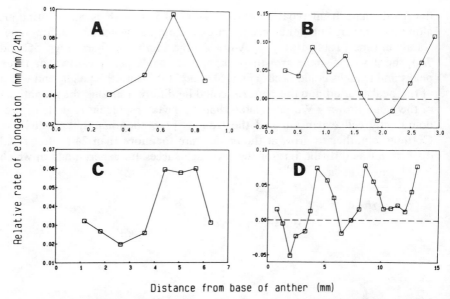

FIG. 7. Growth profiles of individual lily anthers from surface marking experiments. Initial anther lengths (mm): A, 1.11; B, 3.29; C, 6.82; D, 13.88. Figure reproduced from Gould and Lord (1988) with permission from Springer-Verlag.

growth was monitored over 3 d, we detected a pattern that may be interpreted as a basipetally moving waveform, though more data are necessary to confirm this (14).

The patterns of mitotic activity mirror those of the surface growth profiles (Figs. 8, A-D), even with respect to lack of synchrony in peak location in the anthers of a single flower. The large anthers showed more than one mitotic peak (Fig. 8, C and D). Since the mitotic patterns of anthers undamaged prior to fixation recapitulate those of the growth analysis, the wounding that resulted from the marking experiments was probably not responsible for the fact that the spatial distribution of growth along an anther is continually changing in time.

Mitosis in the anther ceases at 6 to 9 mm, meiosis in the pollen mother cells occurs between 9 to 12 mm, and all growth subsequent to this is by cell expansion alone. Anthers larger than 9 mm showed significant differences in cell length along the epidermis (Fig. 9, A-D), which explains the continuity of the waveform even after cell division has ceased in the anther. At this point in anther development, the growth peaks correspond solely to localized regions of cell expansion.

We propose that the waveform, since it exists from the initial stages of anther growth to maturity, is coordinating all aspects of anther morphogenesis. These waves move very slowly (~ 3 mm/24 h), and the peaks may be broad enough to nearly encompass the anther in later stages of maturation. There are no obvious gradients in cellular differentiation along the

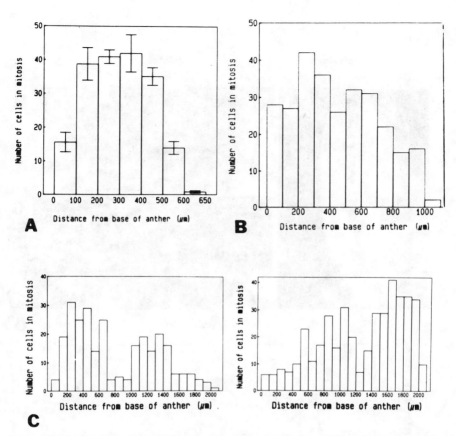

FIG. 8. Frequency of cells in mitosis in 100 μm stretches of lily anthers. Data taken from stained serial sections of anthers. A, 0.65 mm tall anthers. Data are the mean values of five anthers from one bud. Bars denote ± 1 S.E. B, 1.1 mm tall anthers. C, D, Two anthers from the same bud, each 2.1 mm tall. Figure reproduced from Gould and Lord (1988) with permission from Springer-Verlag.

anthers, as many researchers have noticed (7, 17), but detailed studies of chromosomal activity during meiosis have demonstrated subtle gradients that may correspond to the wave pattern we have observed (21, 22).

Summary

The surface growth analysis allowed us to detect a growth pattern not previously described in determinate plant organs. There appear to be factors controlling the growth pattern which can then modify the spatial distribution of mitosis; a supra-cellular phenomenon that lends credence to de Bary's view that "the plant forms cells, not cells the plant" (2). The waveform in the anther persists both during the phase of cell division and long after the last cell has divided. In the tepal, the predominantly basal growth (phase II) is not simply the result of a basal intercalary "meristem," since its onset precedes

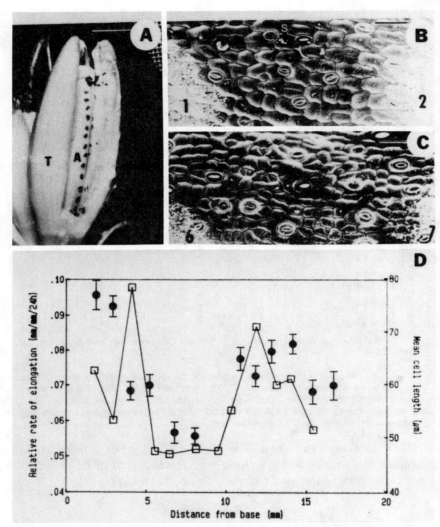

FIG. 9. Marked anther, 17.4 mm long. A, Appearance *in situ*. B, C, SEM images of cryo-preserved anther in A. B, Portion between 1st and 2nd mark from base. C, Portion between 6th and 7th mark from base. Granular areas are the carbon marks, labelled according to position from the base of the anther. Note differences in cell size. D, Growth profile □ , and mean cell lengths ● of each region of the anther. Bars denote ± 1 S.E. Epidermal cell length varied significantly among different regions of the anther (analysis of variance; $p < 0.01$). A = anther; S = stomate; T = tepal. Bar: A, 5 mm; B, C, 0.1 mm. Figure reproduced from Gould and Lord (1988) with permission from Springer-Verlag.

the shift of mitosis to the base and its duration outlasts cell division altogether. The evidence suggests that there are factors controlling the growth pattern which can then modify the spatial distribution of mitosis. Similar studies on primordial stages of leaf development are too few to make generalizations (5), but it is possible that the wave pattern we see in early tepal growth and throughout development of the anther characterizes early stages of organogenesis in leaves. Poethig and Sussex describe what appears to be a wave of cell division propagating along a 6 mm tobacco leaf, but soon thereafter, mitoses cease in a basipetal pattern typical of leaves (18). Preliminary work on the gynoecium demonstrates initially nonsteady growth as seen in the phase I tepals and throughout anther growth (see Lord, Eckard, and Crone abstract this volume, p. 190).

A possible mechanism for the waveform is pulses of a growth promoter (or inhibitor) propagating down the anther or peaking randomly in the phase I tepal. In phase II tepals, the chemical could be restricted to the base, and then in phase III move acropetally as a pulse and effect separation of the tepals in anthesis. Alternatively, mechanical forces generated by local peaks of cell division might, once set up, perpetuate the wave pattern in the anthers. In animal systems, others have proposed that growth is generated by a wavelike propagation of signals from "localized pacemakers" and that a molecular approach to the study of embryogenesis is premature until we understand the nature of these driving forces (13).

Growth in the tepals and anthers of lily is nonsteady, with growth centers constantly shifting; the growth pattern itself appears to modify the spatial distribution of mitosis. Less damaging means of marking the surfaces of growing organs are necessary, though, before attempting similar studies on primordial stages of leaf development and smaller flowers.

Acknowledgments--The authors wish to thank J. P. Hill and K. J. Eckard for their editorial comments. This work was supported by NSF grant PCM-8512062.

LITERATURE CITED

1. ARBER A 1937 The interpretation of the flower: a study of some aspects of morphological thought. Biol Rev 12: 157-184
2. BARLOW PW 1982 Root development. *In*: HS Smith, D Grierson, eds, The Molecular Biology of Plant Development. University of California Press, Berkeley, CA, pp 185-222
3. BOKE NH 1948 Development of the perianth in *Vinca rosea* L. Am J Bot 35: 413-423
4. BOKE NH 1949 Development of the stamens and carpels in *Vinca rosea* L. Am J Bot 36: 535-547
5. CHEN JCW 1963 Quantitative analyses of the growth of leaf primordia in *Eupatorium rugosum* Houtt. PhD Diss, University of Pennsylvania

6. EARLE E 1982 Gametogenesis, fertilization and embryo development. *In* H Smith, D Grierson, eds, The Molecular Biology of Plant Development. University of California Press, Berkeley, CA
7. ERICKSON RO 1948 Cytological and growth correlations in the flower bud and anther of *Lilium longiflorum*. Am J Bot 35: 729-739
8. ERICKSON RO, WK SILK 1980 The kinematics of plant growth. Sci Am 242: 134-151
9. ESAU K 1965 Plant Anatomy. John Wiley and Sons, New York
10. FAHN A 1982 Plant Anatomy. Pergamon Press, New York
11. FOSTER AS, EM GIFFORD JR 1974 Comparative Morphology of Vascular Plants. WH Freeman and Co, San Francisco
12. GOLDBERG RB 1986 Regulation of plant gene expression. Phil Trans R Soc London B 314: 343-353
13. GOODWIN BC, MH COHEN 1969 A phase-shift model for the spatial and temporal organization of developing systems. J Theor Biol 25: 49-107
14. GOULD KS, EM LORD 1988 Growth of anthers in *Lilium longiflorum*: a kinematic analysis. Planta 173: 161-171
15. GOULD KS, EM LORD 1989 A kinematic analysis of tepal growth in *Lilium longiflorum*: . Planta (in press)
16. GREEN PB 1988 A theory for inflorescence development and flower formation based on morphological and biophysical analyses in *Echeveria*. Planta 175: 153-169
17. HOTTA Y, H STERN 1963 Molecular facets of mitotic regulation I. Synthesis of thymidine kinase. Proc Natl Acad Sci USA 49: 648-654
18. POETHIG RS, IM SUSSEX 1985 The cellular parameters of leaf development in tobacco: a clonal analysis. Planta 165: 170-184
19. RITTERBUSCH A 1980 The spatio-temporal patterns of growth and development in floral ontogenesis as visualized by Bildscharen and trajectories. Flora 169: 405-423
20. SILK WK 1984 Quantitative descriptions of development. Annu Rev Plant Physiol 35: 479-518
21. WALTERS MS 1976 Variation in preleptotene chromosome contraction among three cultivars of *Lilium longiflorum*. Chromosoma 57: 51-80
22. WALTERS MS 1980 Premeiosis and meiosis in *Lilium* "Enchantment". Chromosoma 80: 119-146

ENVIRONMENTAL AND CHEMICAL CONTROLS OF FLOWER DEVELOPMENT

JEAN-MARIE KINET

Centre de Physiologie Végétale Appliquée (I.R.S.I.A.), Département de Botanique, Université de Liége, Sart Tilman, B-4000 Liége, Belgium

Flower and inflorescence development are complex processes, extending over a protracted period. Numerous events are coordinated in time and space to produce, in harmonious fashion, many different structures. Elucidation of the mechanisms involved in the control of reproductive development is strongly dependent on identifying, spatially localizing, and temporally ordering these events. This work requires a strict developmental control which allows synchronization of plants and, consequently, the detection of short-term events and the establishment of their sequence.

Fortunately, flower development in many species is strongly influenced by environmental factors, such as light and temperature, or by application of chemicals. In some instances, the control is absolute. My aim in this paper will be to show that, using such controlling factors, it is possible to detect stages essential to reproductive development. Tomato (*Lycopersicon esculentum* Mill.) will serve as a model plant. Its inflorescence is a cyme, which is initiated by the apical meristem and develops laterally on the stem after being displaced from its terminal position by the active growth of the adjacent bud in the axil of the last-formed leaf. This process is repeated for each sympodial unit. All the work reported here was done on the first inflorescence. In experiments on late stages of reproductive development, synchronization of plants was reset before experimentation by a careful selection of individuals all at the same stage of floral development, *i.e.* the macroscopic appearance of the inflorescence.

ENVIRONMENTAL CONTROL

Control by Light. Light is one of the major environmental factors controlling flower development. Both daylength and irradiance, either independently or in combination, exert a profound influence.

In day-neutral species, like tomato, flower development can often be controlled by the manipulation of the daily light integral. We have designed three different experimental systems leading to either the complete development of the inflorescence up to flower opening or the total or partial failure of the reproductive structure (Fig. 1). When plants are grown under constant favorable light conditions, consisting of 16- to 20-h LD given at a photon flux density of 180 μmol.m^{-2}.s^{-1} at the top of the canopy (case 2 of Fig. 1), floral transition occurs during the 4th week, and the inflorescence appears macroscopically around the 45th day. Anthesis of the first flower of the inflorescence is reached about 2 weeks later, *i.e.* 2 months after sowing. In these favorable light conditions, some of the last initiated flower buds invariably abort.

Under constant unfavorable light conditions, which consisted of 8-h SD with a photon flux density of 90 μmol.m^{-2}.s^{-1} (case 1 of Fig. 1), floral

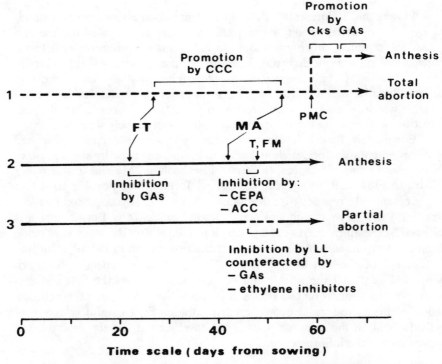

FIG. 1. Diagram summarizing the temporal interactions between light conditions and PGRs for the control of inflorescence development in the tomato. Broken lines, unfavorable light conditions (LL); solid lines, favorable light conditions; FT, floral transition; MA, macroscopic appearance of the inflorescence; PMC, pollen mother cell stage; T,FM, tetrad-free microspore stage [computed from Abdul *et al.* (1), Kinet *et al.* (8), Kinet and El Alaoui Hachimi (5)].

initiation, as well as macroscopic appearance of the inflorescence, are later than under high light, indicating that the rate of growth of the reproductive structure is slowed down. But the essential point is that development never proceeds to anthesis and failure of the whole inflorescence invariably occurs in all individuals (total abortion). In the first flower of the aborting inflorescence, all floral appendages are present, but mitotic activity and nuclear DNA synthesis are completely stopped. In the anthers, the cells of the sporogenous tissue are halted at a premeiotic stage and, in the ovules, development never goes beyond the early differentiation of the archesporial cell (9).

However, unfavorable light conditions are not permanently required from sowing to cause flower failure in tomato. In plants grown under favorable light conditions, there is a relatively narrow time dependence for low-light induced abortion (case 3 of Fig. 1). The critical stage is between 5 to 6 and 10 to 12 d after the macroscopic appearance of the inflorescence, when sporogenesis is in progress in the anthers of the first flower of the inflorescence, at the tetrad and free microspore stages (5). Unfavorable light conditions at that time cause partial abortion, *i.e.* failure of the whole inflorescence in only a part of the individuals and, in plants which succeed in developing flowers to anthesis, abortion of individual floral buds at unusual positions.

In contrast to tomato, many species have a marked photoperiodic requirement for reproductive development, either facultative when they develop their flowers in all photoperiods but faster under a favorable daylength, or absolute when flower buds abort prematurely under inappropriate photoperiodic conditions (8).

Control by Temperature. One of the most obvious and universal effects of temperature is upon the rate of development of the reproductive structures. As a rule, the higher the temperature, the earlier the opening of the flowers, provided that premature failure does not occur since abortion may also be triggered by increasing temperature (8).

Interactions with light are evident in tomato where the detrimental effect of a short stay under insufficient light during sporogenesis is totally counteracted by decreasing the temperature by $4^\circ C$ only (3). As in many other plants, the response to one environmental parameter is, thus, dependent on other factors, and an accurate manipulation of the flowering behavior implies the adjustment of all environmental parameters to an adequate level.

Temperature also has some interesting morphogenetic effects. In tomato, also in tulip, rose, carnation, etc., there is a tendency to increase the number of floral parts in the flower in response to exposure to low temperatures (8, 25). The number of flowers developing per inflorescence in tomato is also higher after chilling (12, 20). This increased flower number seems to result from the transformation of a floral primordium into a new reproductive axis meristem inducing the inflorescence to branch. Low temperature is most effective during floral transition (Fig. 2).

CORRELATIVE INFLUENCES

All plant parts intimately interplay for the control of reproductive development. Flower number in the tomato inflorescence is increased by reducing root temperature; this effect is graft transmissible, suggesting that a root-generated signal is moving up to the differentiating inflorescence where it modifies its structure (20). Apical dominance has also been implicated in tomato since light conditions, decapitation, and growth regulators affect development of both lateral shoots and the inflorescence in the same manner (4, 15).

Besides these long-distance interactions between the reproductive and vegetative plant organs, short-distance interactions inside the inflorescence and flower primordia also exist. When a first flower reaches anthesis in the tomato inflorescence, development of the other flower buds in this inflorescence is influenced as demonstrated by the observation that they become less susceptible to low light and that growth correlations between their different organs are modified: the rate of growth of petals, stamens, and style, all deciduous organs with determinate growth, is increased as compared to that of sepals (8). The mechanism responsible for this stimulation is unknown, but these results reveal that the fate

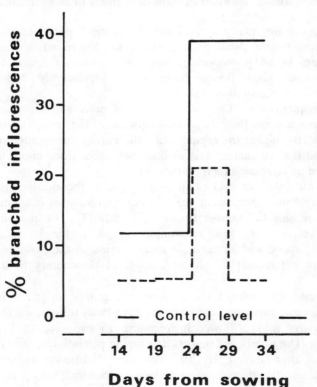

FIG. 2. Effects of a 5-d (broken lines) or a 10-d (solid lines) treatment at 10°C, given at different ages, upon the branching of the first inflorescence of tomato plants otherwise grown at 20°C. Floral transition in control untreated plants occurred between the 24th and 29th day.

of the different flower buds of the cyme is at least partly linked to the destiny of the neighboring flowers and that the response of individual buds to the environment may be modulated by positional influences. Interestingly, the relative growth of the ovary to that of the sepals remains unchanged after the first anthesis, suggesting that these organs share some common features and that ovary development is related to sepal development while petal development would be linked to stamen development. This hypothesis was formulated by Nitsch (17) and accounts for observations on many other species as well (8, 13).

CHEMICAL CONTROL

Correlative influences imply the existence of signals, presumably plant growth regulators (PGRs), circulating within the plant. The participation of PGRs in the control of reproductive development in tomato has been demonstrated using the three experimental systems summarized in Figure 1.

In Constant Unfavorable Light Conditions (Case 1). Treating the exclusive inflorescence first with N^6-benzylaminopurine (BA) and then with GA_{4+7} prevents abortion and triggers development up to flower opening (6). This response is not observed when each of the two PGRs is applied alone. GAs alone cause the rapid abscission of the young flower buds after producing a limited elongation of the inflorescence peduncle. The cytokinin (CK) alone stimulates the growth of the inflorescence but, generally, not up to anthesis. Complete development is not observed when the GAs are applied before the CK, showing that the sequential action of the two PGRs is obligate. However, anthesis can be reached with repeated applications of a mixture of BA and GA_{4+7}.

The observation that inflorescences aborting in permanent unfavorable light conditions suffer a severe CK deficiency, in comparison with levels in inflorescences which develop normally in favorable light conditions (10), further supports the view that CKs play a major role in the control of early reproductive development. GAs are not a limiting factor at that time: their level is high in inflorescences undergoing early abortion (10), and the available evidence suggests that high GA activity is inhibitory during first stages of inflorescence development. Indeed, application of the growth retardant 2-chloroethyltrimethylammonium chloride (CCC) from floral transition to macroscopic appearance of the inflorescence results in a reduction of the incidence of abortion induced by high temperature and low light. Concomitantly, CCC causes a decreased yield of diffusible GAs from plant shoot tips (1).

In Constant Favorable Light Conditions (Case 2). When applied at the time of inflorescence initiation, GAs cause the abortion of many flowers, but the development of those inflorescences which avoid abortion is accelerated, resulting in advanced anthesis (8). Several of these inflorescences have increased flower numbers as a consequence of inflorescence branching which, in one experiment, occurred in 80% of the individuals as compared to 10% in control inflorescences. CKs have a similar effect upon inflorescence ramification. Leaf-like structures and leafy shoots are also frequently (70%) produced in the inflorescence of plants treated by GAs at the time of floral transition, and less often after BA applications (30%); they are not detected in untreated plants.

The question as to whether temperature action on inflorescence ramification in tomato is related to changes in PGR status is unsolved. Yet, branching is apparently not dependent on a specific stimulus. These findings reveal the great plasticity of the tomato cyme and suggest that, at initiation, a flower primordium is not determined. For a time, the inflorescence is thus a mosaic composed of a variety of florally determined and undetermined parts.

During the later stages of inflorescence development, *i.e.* after its macroscopic appearance, GAs are without effect upon rate of growth, but they reduce the proportion of flowers that naturally abort indicating a promotive effect (4). GA requirement for late development of flower buds in tomato is also demonstrated in studies using mutants. In the single gene recessive *ga-2* mutant, where no GA-like activity is detectable in extracts from either shoots or fruits (28), flower buds are initiated but do not develop to maturity (16). Their corolla and stamens do not elongate, and they abort before meiosis of the micro- and megasporocytes. The deficiency can be overcome by the application of GA_1 and fertility is restored as it is also in the single gene recessive, stamenless-2 (*sl-2/sl-2*) mutant which is structurally and functionally male sterile (26, 27). Extractable GAs in shoots of this mutant are less than 20% of those in wild-type shoots, and the levels in the floral parts are also reduced (24). In several other species, but perhaps not in cereals (19, 22), late stages of flower development seem to be dependent on GAs (8, 19).

2-Chloroethylphosphonic acid (CEPA), which releases ethylene in plant tissues, and 1-aminocyclopropane-1-carboxylic acid (ACC), the immediate ethylene precursor, inhibit flower and inflorescence development in tomato at the time GAs are promotive and unfavorable light conditions are most detrimental (5).

During a Short Stay in Low Light Conditions (Case 3). GAs and ethylene inhibitors, such as silverthiosulfate (STS) and aminoethoxyvinyl glycine (AVG), counteract the adverse effect of a short stay in low light during meiosis (4, 5). In contrast, CKs are ineffective in these conditions. These findings suggest that the inhibition by low irradiance during sporogenesis is mediated by the antagonist action of GAs and ethylene.

THE MECHANISM OF ACTION OF LIGHT AND PGRs

There is little question that higher photosynthetic activity in the source leaves of tomato is a major contributing factor to high light-induced promotion of reproductive development (7, 15, 23). However, the observations that PGRs are able to mimic light treatments and that the light regime influences the endogenous contents of CKs and GAs in the inflorescence suggest that the light effects are also mediated by PGRs.

Several changes occurring in response to a BA+GA application on inflorescences targeted for abortion in unfavorable light conditions have been recorded (4, 9, 11). An early effect of the regulators is to stimulate cellular activity in the ovules of the first flower of the inflorescence (9). This resumption of the cellular activity detected around the 16th h precedes the stimulation of acid invertase (EC 3.2.1.26) activity (Fig. 3), which is viewed by several authors as a key enzyme in the control of sugar transfer from the phloem to the sink and of assimilate distribution within the plant (2, 14). This result suggests that the increased import of ^{14}C-assimilates found at the 27th h in BA+GA-treated inflorescences (11), although essential, is indirect. PGRs, particularly the CK which exerts its action first, seem to act primarily as cell-division-mediating factors.

The interpretation of the dual effect of GAs as a function of stage of development remains unclear. The promotion of early inflorescence development in tomato plants treated with growth retardants, which inhibit GA biosynthesis, could result from shoot growth limitation and diversion of metabolites for the benefit of

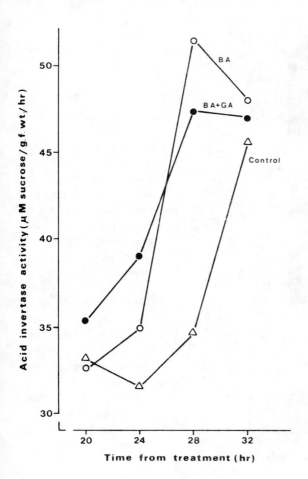

FIG. 3. Changes in acid invertase activity in the first inflorescence of plants continuously grown in unfavorable light conditions, following the application of BA (4.5×10^{-5} M) or a mixture of BA (4.5×10^{-5} M) + GA$_{4+7}$ (6.8×10^{-5} M). The PGRs were applied once, directly to the inflorescence, 6 d after its macroscopic appearance.

the reproductive structure (18). On the contrary, the late stimulatory action of GAs does not seem to be related to assimilate distribution. At that time (from 8 to 2 d before anthesis), sink activity of the inflorescence is markedly decreasing, while it remains almost unchanged in other plant parts (Fig. 4), suggesting that floral development is less dependent on assimilate supply. Furthermore, CKs which are highly effective in diverting assimilates, cannot substitute for GAs in this effect. Hence, the late promotive action of GAs seems to be specific and is possibly essential for meiosis as suggested by studies with various tomato mutants (16, 27). This developmental stage is also under the inhibitory control of ethylene. Sensitivity of microsporogenesis to this gaseous regulator has been found in many other species (5, 8).

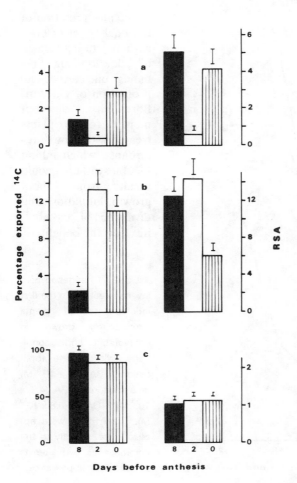

FIG. 4. The partition of the ^{14}C-assimilates translocated out of a source leaf between (a) the first inflorescence, (b) the apical shoot above the inflorescence and (c) the rest of the plant, at three different times during reproductive development. Results are expressed as percent distribution of ^{14}C and relative specific activity (RSA) values. RSA provides a measure of the competitive sink ability of the considered plant part. Vertical bars indicate standard error of the mean. [See Léonard et al. (11) for technical details.]

CONCLUSIONS

This study shows that flower development in tomato can be rather strictly controlled. The experimental evidence indicates that, depending on conditions, this physiological process may be stopped at different stages before and during sporogenesis, a critical step that frequently constitutes the point of no return after which environmental parameters are ineffective in preventing the flower to reach anthesis. The stage at which floral development is inhibited also varies, in different species, according to the position of the flower bud in the inflorescence or on the plant (8). In tomato, where flowers are initiated and develop sequentially in the cymous inflorescence, it is clear that young buds senesce at an earlier stage than the first oldest flower, when abortion is total.

The disruption of the sequence of events, which give rise to normal development of the reproductive structure up to anthesis, by PGRs and the manipulation of the environmental conditions has provided information useful in understanding the complexity of the mechanisms involved in the control of this important physiological step. Clearly, this information is still very fragmentary, and a lack of fundamental work on this highly intricate process prevents advancement. Mutants of flower and inflorescence development are numerous in tomato (21) but, so far, they were not intensively used in morphological, physiological, biochemical, and genetical approaches which would increase the chance of progress.

Acknowledgments--The support of the FRFC of Belgium (Grant no. 2.9009.87) and the University of Liége (Action de Recherche Concertée 88/93-29) is gratefully acknowledged.

LITERATURE CITED

1. ABDUL KS, AE CANHAM, GP HARRIS 1978 Effects of CCC on the formation and abortion of flowers in the first inflorescence of tomato (*Lycopersicon esculentum* Mill). Ann Bot 42: 617-625
2. HO LC, DA BAKER 1982 Regulation of loading in transport systems. Physiol Plant 56: 225-230
3. KINET JM 1982 Un abaissement de la température et une élévation de la teneur en CO_2 de l'atmosphère réduisent l'avortement des inflorescences de tomates cultivées en conditions d'éclairement hivernal. Rev Agric 35: 1767-1772
4. KINET JM 1987 Inflorescence development in tomato: control by light, growth regulators, and apical dominance. Plant Physiol (Life Sci Adv) 6: 121-127
5. KINET JM, H EL ALAOUI HACHIMI 1988 Effects of ethephon, 1-aminocyclopropane-1-carboxylic acid, and ethylene inhibitors on flower and inflorescence development in tomato. J Plant Physiol 133: 550-554
6. KINET JM, D HURDEBISE, A PARMENTIER, R STAINIER 1978 Promotion of inflorescence development by growth substance treatments to tomato plants grown in insufficient light conditions. J Am Soc Hort Sci 103: 724-729

7. KINET JM, RM SACHS 1984 Light and flower development. *In* D Vince-Prue, B Thomas, RE Cockshull, eds, Light and the Flowering Process. Academic Press, London, pp 211-225
8. KINET JM, RM SACHS, G BERNIER 1985 The Physiology of Flowering, Vol. III. CRC Press, Boca Raton, FL
9. KINET JM, V ZUNE, C LINOTTE, A JACQMARD, G BERNIER 1985 Resumption of cellular activity induced by cytokinin and gibberellin treatments in tomato flowers targeted for abortion in unfavorable light conditions. Physiol Plant 64: 67-73
10. LEONARD M, JM KINET 1982 Endogenous cytokinin and gibberellin levels in relation to inflorescence development in tomato. Ann Bot 50: 127-130
11. LEONARD M, JM KINET, M BODSON, G BERNIER 1983 Enhanced inflorescence development in tomato by growth substance treatments in relation to ^{14}C-assimilate distribution. Physiol Plant 57: 85-89
12. LEWIS D 1953 Some factors affecting flower production in the tomato. J Hort Sci 28: 207-220
13. LORD E 1981 Cleistogamy: a tool for the study of floral morphogenesis, function and evolution. Bot Rev 47: 421-449
14. MORRIS DA 1983 Hormonal regulation of assimilate partition: possible mediation by invertase. News Bull, British Plant Growth Regulator Group 6: 23-35
15. MORRIS DA, AJ NEWELL 1987 The regulation of assimilate partition and inflorescence development in the tomato. *In* JG Atherton, ed, Manipulation of Flowering. Butterworths, London, pp 379-391
16. NESTER JE, JAD ZEEVAART 1988 Flower development in normal tomato and a gibberellin-deficient (*ga-2*) mutant. Am J Bot 75: 45-55
17. NITSCH JP 1965 Physiology of flower and fruit development. *In* W Ruhland, ed, Encyclopedia of Plant Physiology, Vol. 15 (Part 1). Springer-Verlag, Berlin, pp 1537-1647
18. NOURAI AHA, GP HARRIS 1983 Effects of growth retardants on inflorescence development in tomato. Scientia Hortic 20: 341-348
19. PHARIS RP, RW KING 1985 Gibberellins and reproductive development in seed plants. Annu Rev Plant Physiol 36: 517-568
20. PHATAK SC, SH WITTWER, FG TEUBNER 1966 Top and root temperature effects on tomato flowering. Proc Am Soc Hort Sci 88: 527-531
21. RICK CM, J ROBINSON 1951 Inherited defects of floral structure affecting fruitfulness in *Lycopersicon esculentum*. Am J Bot 38:639-652
22. ROOD SB, DM BRUNS, SJ SMIENK 1988 Gibberellins and sorghum development. Can J Bot 66: 1101-1106
23. RUSSELL CR, DA MORRIS 1982 Invertase activity, soluble carbohydrates and inflorescence development in the tomato (*Lycopersicon esculentum* Mill.). Ann Bot 49: 89-98
24. SAWHNEY VK 1974 Morphogenesis of the stamenless-2 mutant in tomato. III. Relative levels of gibberellins in the normal and the mutant plants. J Exp Bot 25: 1004-1009

25. SAWHNEY VK 1983 The role of temperature and its relationship with gibberellic acid in the development of floral organs of tomato (*Lycopersicon esculentum*). Can J Bot 61: 1258-1265
26. SAWHNEY VK, RI GREYSON 1973 Morphogenesis of the stamenless-2 mutant in tomato. I. Comparative description of the flowers and ontogeny of stamens in the normal and mutant plants. Am J Bot 60:514-523
27. SAWHNEY VK, RI GREYSON 1973 Morphogenesis of the stamenless-2 mutant in tomato. II. Modifications of sex organs in the mutant and normal flowers by plant hormones. Can J Bot 51: 2473-2479
28. ZEEVAART JAD 1984 Gibberellins in single gene dwarf mutants of tomato. Plant Physiol S75: 186

Plant Reproduction: From Floral Induction to Pollination, *Elizabeth Lord and Georges Bernier*, Eds, 1989, The American Society of Plant Physiologists Symposium Series, Vol. I

A MOLECULAR APPROACH TO FLOWER DEVELOPMENT IN *ARABIDOPSIS*

ELLIOT M. MEYEROWITZ, JOHN L. BOWMAN, HONG MA, USHA VIJAYRAGHAVAN, AND MARTY YANOFSKY

Division of Biology 156-29, California Institute of Technology, Pasadena, CA 91125, USA

Mutations that cause abnormal pattern formation in the development of flowers have been known for centuries. We are using several such mutations in genetic and molecular studies of flower development in the small mustard, *Arabidopsis thaliana*.

Four genes have been genetically analyzed, and the development of flowers of plants mutant in these genes studied in detail. The first of the mutations is *agamous*, which causes petals to develop where stamens are normally found, and a new flower to develop from the cells that would, in wild-type flowers, differentiate into the pistil. The second is *apetala2*, which in some alleles causes stamens to develop where petals are usually found, and leaves to differentiate where sepals are expected to be. Other alleles eliminate the petal and stamen whorls, and cause carpels to develop where two of the sepals are ordinarily found. The third mutation is *pistillata*, which eliminates the stamen whorl, causes the organs of what would normally be the petal whorl to differentiate as sepals, and affects gynoecium development as well. The last mutation is *apetala3*, which causes the organs of what is in wild type the petal whorl to differentiate as sepals, and the organs of what is ordinarily the stamen whorl to develop as carpels.

Some information on the action of these genes has been obtained from production of double-mutant strains, and study of the phenotype of their flowers. In addition, one allele of *apetala2* and the only known allele of *apetala3* are temperature sensitive. This has allowed temperature shift experiments that have shown that the wild-type *apetala2* product is only necessary for a short time very early in flower development, before any organ primordia (other than perhaps the sepal primordia) have appeared or differentiated. The *apetala3* product is required later, at the time of organ differentiation.

Experiments to clone the genes are in progress. A dense RFLP map of the *Arabidopsis* genome has been made, and chromosome walks have been started from nearby DNA segments toward *apetala2* and *pistillata*. A happenstance insertion of a T-DNA element into the *agamous* locus has been detected by Dr. Ken Feldmann of Dupont and, in collaboration, we are using this insertion to clone *agamous*.

GENES EXPRESSED DURING POLLEN DEVELOPMENT

JOSEPH P. MASCARENHAS, DOUGLAS A. HAMILTON, AND DAVID M. BASHE

Department of Biological Sciences, State University of New York at Albany, Albany, NY 12222, USA

The major characteristic that distinguishes angiosperms from other groups of plants is the flower. The primary purpose of the flower, often forgotten in discussions of flower development, is the production of the male and female gametophytes and gametes, and the facilitation of the union of the gametes in fertilization.

The early development of the male gametophyte (pollen) occurs in a young flower bud within the anther after meiosis is complete. The microspores, on release from the tetrad, increase rapidly in size and undergo a change in shape. Microspore mitosis occurs after a long interphase of about 5 d in length in *Tradescantia paludosa* (7) and 7 to 9 d in corn (2). Microspore mitosis results in a vegetative cell which contains the bulk of the cytoplasm of the microspore and a generative cell which inherits a very small amount of cytoplasm and is surrounded by the cytoplasm of the vegetative cell. In the pollen of several plant species, corn for example, the generative cell divides in the pollen grain producing two sperm cells. In most plants however, including *Tradescantia*, the generative cell completes its division after the pollen grains are released from the anther and during pollen tube growth in the style. In *Tradescantia*, the period from microspore mitosis to anthesis is about 4 d (7), and in corn it is 6 to 7 d (2). Following anther dehiscence, the pollen grain is deposited on the stigma of the pistil in a flower where it begins another phase of its development by germinating and producing a pollen tube which grows down into the style. This phase of development is unlike the type of development prior to anthesis and consists largely of tube cell wall synthesis and the synthesis of cell membrane to accompany the rapid increase in length of the tube (4). To reach the embryo sac, pollen tubes grow to appreciable lengths within relatively short periods of time. The style is about 4 to 6 mm in length in *Tradescantia* and, in corn, the silks can reach lengths of 40 to 50 cm. Thus, although the development of the male gametophyte is relatively simple in comparison to the sporophyte, there are a number of discrete events and types of differentiation that occur during its life. For a number of years, we have been interested in the regulation of various molecular events during the development of the male gametophyte. Recent and current work from our laboratory concerning the numbers, types, pattern of expression, and characterization of genes expressed in the male gametophyte will be summarized here.

NUMBERS OF GENES EXPRESSED IN POLLEN AND POLLEN-SPECIFIC GENES

The mature pollen grains of several plant species contain mRNAs that were synthesized prior to anthesis (1, 6, 14). These mRNAs have been isolated and, in cell free translation systems, code for polypeptides that are similar to those synthesized during pollen germination and pollen tube growth (1, 6). In many pollens, these presynthesized mRNAs appear to be required during pollen germination and early tube growth (4).

What is the number of different genes that are active in pollen development which are responsible for the mRNAs present in the mature pollen grain? An analysis of the kinetics of hybridization of cDNA with poly(A)+RNA in excess has been carried out with poly(A)+RNA isolated from mature pollen of *Tradescantia paludosa* (16) and maize (15). These data are summarized in Table I and show that the mRNAs in mature pollen consist of three abundance classes. The first class which is a relatively small fraction of the total poly(A)+RNA of the pollen grains, is low in complexity, and consists of a few different mRNAs which are extremely prevalent, each being present in 26,000 and 32,000 copies per pollen grain of *T. paludosa* and maize, respectively. The mRNAs of the second class, which make up the bulk of the mRNAs, are intermediate in number and in reiteration frequency. The most complex class of mRNAs is made up of 17,000 to 18,000 different sequences, each present on an average in 100 to 200 copies per pollen grain. The total complexity of the pollen mRNAs in both *T. paludosa* and maize is very similar, being in excess of 2×10^7 nucleotides. The mRNAs in the mature

Table I. *Summary of homologous cDNA-poly(A)+RNA hybridizations from pollen of Tradescantia paludosa and maize.*

Component	Fraction of total poly(A)+RNA		Complexity (nucleotides)		No. of diverse mRNAs		No. of copies per sequence per pollen grain	
	T. paludosa	maize	*T. paludosa*	maize	*T. paludosa*	maize	*T. paludosa*	maize
1	0.15	0.35	5.2×10^4	2.1×10^5	44	245	26,000	32,000
2	0.61	0.49	1.6×10^6	6.4×10^6	1,400	6,260	3,400	1,700
3	0.24	0.15	2.1×10^7	1.8×10^7	18,000	17,250	100	195
Total	1.00	1.00	2.3×10^7	2.5×10^7	19,444	23,755	-	-

[a]*Data taken from references 15 and 16.*

pollen grains of both plants are the products of approximately 20,000 different genes. It is likely, as is discussed later, that most of the mRNAs present in the mature pollen grain have been synthesized after microspore mitosis. Many genes, however, are also active during early stages of pollen development after meiosis (5), but we have at present no estimates of their numbers or similarity to those expressed later in pollen development. There are thus a large number of genes involved in pollen development.

In comparison, young shoots of both *Tradescantia* and maize contain about 30,000 different mRNAs (15, 16). In general, the mRNAs in pollen are much more prevalent than those in the shoot. Neither in maize nor in *Tradescantia* shoots are there mRNAs that are present in excess of 25,000 copies per cell and the most abundant fraction contains mRNAs that are present in only a few hundred copies per cell. Even the least abundant fraction in pollen contains sequences that are much more abundant (100-200 copies) than in the corresponding fraction in shoots (5-10 copies) (15, 16). This abundance of mRNAs in mature pollen would seem to suggest a requirement for the rapid synthesis of a large quantity of protein product during the terminal stages of pollen maturation and/or germination and early tube growth.

Heterologous RNA C_ot hybridizations, such as pollen cDNA to shoot poly(A)+RNA or shoot cDNA to pollen RNA, indicate that in *T. paludosa* at least 64% of the pollen mRNAs are also represented in the shoot mRNA, whereas a maximum of about 60% of the shoot mRNAS are found in pollen (16). Similarly, in maize, about 65% of the transcripts in pollen are common to those in shoots. Because of various constraints in this type of analysis, the estimate of pollen mRNAs shared with the sporophyte could be in excess of 90% (15).

We have constructed cDNA libraries to poly(A)+RNA from mature pollen of *Tradescantia* and maize. Based on colony hybridizations of a large number of clones from the libraries, using ^{32}P-cDNAs from pollen and vegetative tissues as probes, it is estimated that about 10% and 20% of the total genes expressed in maize and *Tradescantia* pollen, respectively, might be specific to pollen (11). Similar estimates of the numbers of genes expressed in the male gametophyte as compared to the sporophyte have been made for several plants based on comparisons of isozyme profiles (8, 9, 13) and the results are in basic agreement. In maize, about 72% of the isozymes studied were expressed in both pollen and sporophyte, whereas only 6% of the isozymes were pollen specific (9).

In summary, there are a large number of genes expressed during pollen development. Most of these genes are also expressed in vegetative tissues and a small percentage, probably less that 10%, are unique to pollen.

USE OF CLONED LIBRARIES TO STUDY GENE EXPRESSION IN POLLEN

The libraries constructed contain clones that represent mRNAs that are present in mature pollen. Several of these clones, both pollen specific and those shared with the sporophyte, have been used as probes to study the stage of pollen development when the genes first are activated and to study the pattern of accumulation of the mRNAs (11). Such Northern analyses have shown that all the clones that have been studied thus far represent genes that are first activated after microspore mitosis. The mRNAs accumulate thereafter reaching a maximum concentration in the mature pollen grain just prior to anthesis. These libraries therefore, represent genes that are activated relatively late during pollen maturation in the anther (11). In contrast, using a heterologous actin probe, actin mRNA is first detectable in *Tradescantia* in the microspore soon after release from the tetrads. Actin mRNA accumulates thereafter, reaches a maximum concentration at late pollen interphase, and decreases substantially in the pollen grain at anthesis (11). Alcohol dehydrogenase and β-galactosidase are probably other examples of genes that are expressed early after meiosis (10, 12). These results suggest that there must be at least two sets of genes with respect to the timing of their activity during pollen development. The first, which we term "early" genes, represented by actin, become active soon after meiosis; their mRNAs reach their maximum accumulation by late pollen interphase and then decrease substantially by anthesis. One might expect these genes to have primary functions in terms of their proteins, during earlier stages of pollen development. The second class to which most of the cDNA clones in our libraries belong, are activated after microspore mitosis and the mRNAs accumulate reaching a maximum in the mature pollen grain. The pattern of accumulation of these "late" gene mRNAs would seem to suggest a major function for these mRNAs during the latter part of pollen development and/or during pollen germination and tube growth.

CHARACTERIZATION OF A POLLEN-SPECIFIC GENE FROM MAIZE

One of the cDNA clones from the library, Zmc13, has been selected for detailed characterization. This clone is pollen specific. Northern analysis shows that Zmc13 hybridizes to a mRNA of approximately 985 nucleotides in length that is found only in pollen but not in RNA isolated from shoots, roots, kernels, ovules or silks (3). Zmc13 represents a gene that is present in a very few copies in the corn genome (11). The cDNA has been sequenced; it is 929 nucleotides in length and, in addition, has a 47 nucleotide poly(A) tail. Primer extension analysis indicates that Zmc13 is a full-length copy of the mRNA, which codes for a predicted polypeptide that is 170 amino acid residues long and has a mol wt of 18.3 kD. The hydropathy profile strongly suggests a signal peptide at the amino terminus. The function of this protein is not yet known. A computer search of the nucleotide and protein sequence data bases has not revealed any meaningful homology with known proteins (3).

A genomic clone corresponding to Zmc13 has been isolated by screening a genomic library of the inbred maize line W-22. The cDNA clone is colinear with the genomic clone with no introns being present. The genomic clone has been sequenced including greater than 1000 bp 5' to the start of transcription and 200 bp 3' to the termination of transcription. There are some unusual features in both the 5' and 3' flanking regions. Whether these are responsible for pollen specificity of expression will require additional studies.

LOCALIZATION OF Zm13 mRNA

In situ hybridizations using single stranded ^{35}S-labeled riboprobes demonstrate that the mRNA is present in the cytoplasm of the vegetative cell in the mature pollen grain. The silver grains are uniformly distributed throughout the vegetative cell cytoplasm (3). The Zm13 mRNA is also uniformly present throughout the pollen tube cytoplasm after germination. Zm13 mRNA is thus a product of the vegetative cell nucleus rather than the generative cell or sperm cells (3).

Acknowledgements--Supported by National Science Foundation Grant DCB88-10165.

LITERATURE CITED

1. Frankis RC, JP Mascarenhas 1980 Messenger RNA in the ungerminated pollen grain: a direct demonstration of its presence. Ann Bot 45: 595-599
2. Frova C, G Binelli, E Ottaviano 1987 Isozyme and hsp gene expression during male gametophyte development in maize. *In* MC Rattazzi, JG Scandalios, GS Whitt, eds, Isozymes: Current Topics in Biological and Medical Research, Vol 15: Genetics, Development, and Evolution. Alan R Liss, New York, pp 97-120
3. Hanson DD, DA Hamilton, JL Travis, DM Bashe, JP Mascarenhas 1989 Characterization of a pollen specific cDNA clone from *Zea mays* and its expression. The Plant Cell 1 (in press)
4. Mascarenhas JP 1975 The biochemistry of angiosperm pollen development. Bot Rev 41: 259-314
5. Mascarenhas JP 1988 Anther- and pollen-expressed genes. *In* DPS Verma, RB Goldberg, eds, Temporal and Spatial Regulation of Plant Genes. Springer-Verlag, New York, pp 97-114
6. Mascarenhas NT, D Bashe, A Eisenberg, RP Willing, CM Xiao, JP Mascarenhas 1984 Messenger RNAs in corn pollen and protein synthesis during germination and pollen tube growth. Theor Appl Genet 68: 323-326
7. Peddada L, JP Mascarenhas 1975 5S ribosomal RNA synthesis during pollen development. Dev Growth Diff 17: 1-8

8. RAJORA OP, L ZSUFFA 1986 Sporophytic and gametophytic gene expression in *Populus deltoides* Marsh, *P. nigra* L and *P. maximowiczii* Henry. Can J Genet Cytol 28: 476-482
9. SARI-GORLA M, C FROVA, G BINELLI, E OTTAVIANO 1986 The extent of gametophytic-sporophytic gene expression in maize. Theor Appl Genet 72: 42-47
10. SINGH MB, PM O'NEILL, RB KNOX 1985 Initiation of post meiotic beta- galactosidase synthesis during microsporogenesis in oilseed rape. Plant Physiol 77: 229-231
11. STINSON JR, AJ EISENBERG, RP WILLING, ME PE, DD HANSON, JP MASCARENHAS 1987 Genes expressed in the male gametophyte of flowering plants and their isolation. Plant Physiol 83: 442-447
12. STINSON JR, JP MASCARENHAS 1985 Onset of alcohol dehydrogenase synthesis during microsporogenesis in maize. Plant Physiol 77: 222-224
13. TANKSLEY SD, D ZAMIR, CM RICK 1981 Evidence for extensive overlap of sporophytic and gametophytic gene expression in *Lycopersicon esculentum*. Science 213: 454-455
14. TUPY J 1982 Alterations in polyadenylated RNA during pollen maturation and germination. Biol Plant 24: 331-340
15. WILLING RP, D BASHE, JP MASCARENHAS 1988 An analysis of the quantity and diversity of messenger RNAs from pollen and shoots of *Zea mays*. Theor Appl Genet 75: 751-753
16. WILLING RP, JP MASCARENHAS 1984 Analysis of the complexity and diversity of mRNAs from pollen and shoots of *Tradescantia*. Plant Physiol 75: 865-868

REGULATION AND DEVELOPMENT OF MALE STERILITY IN TOMATO AND RAPESEED

V K Sawhney, S K Bhadule, P L Polowick and R Rastogi

*Department of Biology, University of Saskatchewan,
Saskatoon, Saskatchewan, S7N 0W0, Canada*

Male sterility in flowering plants is potentially useful for both the plant breeders and developmental botanists. The former use the male sterile lines in breeding programs, whereas to the latter, such lines are excellent experimental material for investigating the mechanisms involved in pollen and stamen development.

The expression of male sterility can vary. It can range from the complete absence of stamens to the failure of dehiscence of anthers containing normal pollen grains. Further, male sterility may either be nuclear or genic (GMS), or cytoplasmic (CMS), or it may involve both the nuclear and cytoplasmic genes (5, 6). Thus, a wide variety of male sterile phenotypes with different genetic bases are available for experimentation.

In the past, a major emphasis was placed by researchers on investigating the genetics and the development of male sterile lines for hybrid production. Much less is known of the ontogenetic processes in pollen abortion and the experimental control of male sterility by chemical and environmental factors. Such studies are essential for fully understanding the mechanisms involved in male sterility. Also, the ability to experimentally manipulate the expression of male sterility should be advantageous for its utilization in breeding programs.

In a GMS stamenless-2 mutant in tomato (*Lycopersicon esculentum*), and the CMS *ogu* line in rapeseed (*Brassica napus*), we have investigated: (*i*). the structural characteristics associated with the breakdown of microsporogenesis and (*ii*) the regulation of male sterility by temperature and plant growth substances. The following is a brief review of our findings on these two systems.

STAMENLESS-2 MUTANT IN TOMATO

The stamenless-2 is a single gene mutant which, in the homozygous recessive condition (sl-2/sl-2), produces abnormal stamens that do not fuse laterally, are shorter in length, paler in color than the normal (+/+), and produce naked external ovules on the adaxial surface (20). Under the normal growth conditions, the mutant anthers produce microspores but the majority either degenerate or are non-viable (19, 20).

Microsporogenesis. An examination of normal and mutant anthers revealed that until the tetrad stage, there were no differences in the sporogenous tissue of the two lines (19). Subsequently, whereas the pollen development in normal anthers progressed from the binucleate stage to vacuolate stage to mature pollen, most mutant microspores, particularly those in the center of the locule, were either empty, had large vacuoles, or had degenerated (19).

Major differences were noted in the tapetum, the tissue which surrounds and supplies several important substances to the developing microspores (3, 10). First, the tapetal cells in the mutant were more enlarged and were bilayered at places, whereas the tapetum in the normal was single layered (19). Secondly, immediately prior to degeneration, the tapetal cells in the mutant contained very large vacuoles. Thirdly, and most importantly, the tapetum degeneration in the mutant was delayed. In the normal anthers, the tapetal degeneration occurred soon after the release of microspores from tetrads, but in the mutant it occurred at a later stage. Consequently, unlike the normal (Fig. 1), many of the microspores in the mutant were devoid of an exine (Fig. 2). In the absence of a wall, such microspores enlarged (Fig. 3) and eventually degenerated. It was suggested that the abnormal development of the tapetum, and in particular the delay in its degeneration, deprived the microspores of some essential metabolites required for exine deposition (19).

Esterase Activity and Isozymes. Esterases are believed to have a role in pollen development. This suggestion is based on the observations that esterases were localized specifically in the tapetum prior to its degeneration, and on the developing microspores (22-24).

An analysis of esterase isozymes showed that in both the normal and mutant anthers, there were two major isozymes. The first appeared at early stages when the tapetum was intact, and the second at the start of exine deposition (2). Both the isozymes were of greater intensity in the normal anthers with one exception. The first isozyme was of greater intensity in the mutant at the stage when the tapetum was intact in the mutant, but had degenerated in the normal (2). The overall activity of esterase was, however, lower in mutant anthers at all stages of development.

The lower activity of esterases, and particularly the lower intensity of the two major isozymes, suggested that they may be involved in the lack of exine deposition in mutant microspores. However, whether the reduction in esterase activity is the cause or consequence of other processes involved in pollen degeneration could not be established.

Role of Gibberellins (GAs). The role of gibberellins in stamen development has been reported in many angiosperm flowers (4, 7). In the sl-2/sl-2

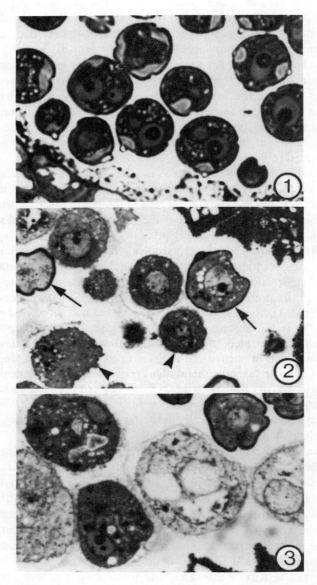

FIG. 1-3. Fig. 1. Cross section of an anther of a normal tomato showing binucleate microspores with a well developed exine, X 1125. Fig. 2. Cross section of an sl-2/sl-2 mutant anther showing the presence of exine on some microspores (arrows), but not on others (arrowheads), X 1320. Fig. 3. Some of the enlarging microspores of the sl-2/sl-2 mutant that are without an exine and have large vacuoles and a thin cytoplasm, X 1320. Reproduced from Sawhney and Bhadula (19) with permission from National Research Council of Canada.

mutant, exogenous applications of GA_3 induced the formation of normal-looking stamens which also contained viable pollen (21). This observation suggested the possibility that the mutant plants may be deficient in endogenous GAs. An analysis of endogenous GAs showed that the vegetative parts and flowers of the mutant contained a lower level of GA-like substances (16).

The role of GAs in stamen development of the mutant flowers was also examined by culturing young floral buds *in vitro*. The normal buds cultured on a Murashige and Skoog medium required the presence of a cytokinin for complete flower development (13). However, the mutant buds grown on the same medium showed limited growth (14), but in the presence of GA_3, they produced flowers with the mutant phenotype (14). Thus, an exogenous supply of GA_3 was essential for the growth of mutant flowers. However, unlike the *in vivo* experiments, normal pollen was not produced in the cultured mutant buds. This suggested that in addition to GAs, another factor, presumably supplied by other parts of the plant, is also required for stamen development.

The role of GAs in stamen development was further examined by experiments with 2-chloroethyltrimethyl ammonium chloride (CCC) and ABA. CCC inhibits GA_3 biosynthesis and ABA counters the gibberellin effect in many systems. Both these substances specifically inhibited stamen growth in normal flowers, and suppressed the differentiation of microspores (15). These observations reinforce the conclusion that gibberellins are essential for stamen development in tomato.

Temperature Regulation. The stamen development in the sl-2/sl-2 mutant is variable under field conditions (20). This was related, in part, to the varying temperature conditions. Mutant plants grown in the intermediate temperatures (23°C day/18°C night) produced mutant characteristics described above. However, if plants were grown under high temperatures (28°C day/23°C night), carpel-like structures were produced in place of stamens (17). In contrast, in mutant plants grown in low temperatures (18°C day/15°C night), the stamens resembled the normal and produced viable pollen (17). Thus the expression of male sterility in the sl-2/sl-2 mutant is strongly influenced by temperature conditions. Whether the restoration of fertility in the mutant at low temperatures is accompanied by changes in endogenous gibberellins is not known.

OGU CMS LINE IN RAPESEED

The *ogu* CMS line was developed by introducing the nucleus of *B. napus* into the male sterile cytoplasm of *Raphanus sativus* (1, 9). This line was backcrossed and maintained into cv Westar of *B. napus*.

Both the long and short stamens of the *ogu* flowers were sterile and the morphology of anthers varied from near-normal stamens to those resembling carpels with external ovules and stigmatic surfaces.

Temperature Regulation. The stamen development in *ogu* flowers was also strongly influenced by temperature conditions. In low temperatures (18°C day/15°C night), CMS stamens resembled the carpels; they had an ovary-like region with internal and external ovules, possessed stigmatic surfaces, and lacked the microsporogenous tissue (12). In intermediate temperatures (23°C day/18°C night), stamens possessed both the carpel and anther features, *i.e.* they produced external ovules and showed some stages of microsporogenesis.

In CMS flowers grown in high temperatures (28°C day/23°C night), normal-looking stamens were produced (12). The surface features of anthers were similar to the normal (11) and microsporogenesis, but not megasporogenesis, was observed in many anthers (12).

Microsporogenesis. The development of microspores was variable in male sterile anthers grown in different temperatures.

As indicated above, in low temperatures, microsporogenesis was not observed in CMS anthers. The external ovules produced, however, did show some stages of megasporogenesis.

Microsporogenesis was observed in intermediate and high temperature grown CMS anthers. In the former, the maximum development was the tetrad stage, whereas in the latter microsporogenesis proceeded up to the microspore stage. In both cases, however, only a small percentage of anthers showed the maximum development.

Wherever microsporogenesis was observed, the development of microspores appeared normal both at the light and electron microscopic levels. However, the microspores showed a poor deposition of exine. The tapetum showed many structural changes different from the normal, *e.g.* the production of large vacuoles, and abnormal mitochondria and ER. In some cases, membrane-bound bodies with parallel striations were observed. The function of these bodies is, however, not known.

CONCLUSION

It is evident that in both the GMS sl-2/sl-2 mutant in tomato and the *ogu* CMS line in rapeseed the expression of male sterility is influenced by temperature conditions, although the response by the two systems was opposite. In the sl-2/sl-2 mutant, low temperatures promoted the formation of normal stamens and high temperatures induced the production of carpel-like structures. These temperatures had a reverse effect on the *ogu* CMS line. In any event, the ability to manipulate stamen development in the two systems is useful for investigations into the mechanisms of male sterility.

In the sl-2/sl-2 mutant, gibberellins appear to have an important role in stamen development. The mutant, unlike the normal, floral buds require GA_3 for growth *in vitro*, contain a lower level of gibberellin-like substances than the normal, and young buds treated with GA_3 produce normal stamens with viable pollen. Further, the inhibition of gibberellin synthesis or its action, by CCC and ABA, respectively, in normal floral buds affected the growth of stamens specifically. Thus, the sl-2/sl-2 allele seems to affect either the biosynthesis of gibberellins, or its metabolism, or both, which in turn affect pollen development.

The analysis of microsporogenesis in both the sl-2/sl-2 mutant and *ogu* line revealed that pollen abortion is associated with major structural changes in the tapetum. The tapetum supplies a variety of substances to the developing microspores (10) and any abnormalities in this tissue would conceivably have an adverse effect on pollen development. Irregularities in the tapetum development have been shown in other GMS and CMS systems (6, 8).

Finally, the ability to regulate stamen development by gibberellins and temperature conditions is useful in breeding programs. In the sl-2/sl-2 mutant particularly, these treatments allow the production of pure male sterile seed that can be used directly for hybrid production (18).

Acknowledgments--These studies were supported by operating grants from the Natural Sciences and Engineering Research Council of Canada.

LITERATURE CITED

1. BANNEROT H, L HOULIDARD, Y CHUPEAU 1977 Unexpected difficulties met with radish cytoplasm in *Brassica oleracea*. Eucarpia Cruciferae Newsletter 2: 16
2. BHADULA SK, VK SAWHNEY 1987 Esterase activity and isozymes during the ontogeny of stamens of male fertile *Lycopersicon esculentum* Mill., a male sterile stamenless-2 mutant and the low temperature-reverted mutant. Plant Sci 52: 187-194
3. BHANDARI NN 1984 The microsporangium. *In* BM Johri, eds, Embryology of Angiosperms. Springer-Verlag, Berlin, pp 53-121
4. CHAILAKHYAN MKH, VN KHRIANIN 1987 Sexuality in Plants and Its Hormonal Regulation. Springer-Verlag, Berlin
5. FRANKEL R, E GALUN 1977 Pollination Mechanisms, Reproduction and Plant Breeding. Springer-Verlag, Berlin
6. KAUL MLH 1988 Male Sterility in Higher Plants. Springer-Verlag, Berlin
7. KINET JM, RM SACHS, G BERNIER 1985 The physiology of flowering. Vol. III. CRC Press, Boca Raton, FL
8. LASER KD, NR LERSTEN 1972 Anatomy and cytology of micro-sporogenesis in cytoplasmic male-sterile angiosperms. Bot Rev 38: 425-454
9. OGURA H 1968 Studies on the new male-sterility in Japanese radish, with special reference to the utilization of this sterility towards the practical raising of hybrid seeds. Mem Fac Agric Kagoshima Univ 6: 39-78
10. PACINI E, GG FRANCHI, M HESSE 1985 The tapetum: its form, function, and possible phylogeny in Embryophyta. Plant Syst Evol 149: 155-185

11. POLOWICK PL, VK SAWHNEY 1986 A scanning electron microscopic study on the initiation and development of floral organs of *Brassica napus* L. (cv Westar). Am J Bot 73: 254-263
12. POLOWICK PL, VK SAWHNEY 1987 A scanning electron microsocpic study on the influence of temperature on the expression of cytoplasmic male sterility in *Brassica napus*. Can J Bot 65: 807-814
13. RASTOGI R, VK SAWHNEY 1986 *In vitro* culture of young floral buds of tomato (*Lycopersicon esculentum* Mill.). Plant Sci 47: 221-227
14. RASTOGI R, VK SAWHNEY 1988 Flower culture of a male sterile stamenless-2 mutant of tomato (*Lycopersicon esculentum*). Am J Bot 75: 513-518
15. RASTOGI R, VK SAWHNEY 1988 Suppression of stamen development by CCC and ABA in tomato floral buds cultured *in vitro*. J Plant Physiol (in press)
16. SAWHNEY VK 1974 Morphogenesis of the stamenless-2 mutant in tomato. III. Relative levels of gibberellins in the normal and mutant plants. J Exp Bot 25: 1004-1009
17. SAWHNEY VK 1983 Temperature control of male-sterility in a tomato mutant. J Hered 74: 51-54
18. SAWHNEY VK 1984 Hormonal and temperature control of male-sterility in a tomato mutant. Proceedings of the VIII International Symposium on Sexual Reproduction in Seed Plants, Ferns and Mosses. PUDOC Publ, Wageningen, pp 36-38
19. SAWHNEY VK, SK BHADULA 1988 Microsporogenesis in the normal and male-sterile stamenless-2 mutant of tomato (*Lycopersicon esculentum*). Can J Bot 66: 2013-2021
20. SAWHNEY VK, RI GREYSON 1973 Morphogenesis of the stamenless-2 mutant in tomato. I. Comparative description of the flowers and ontogeny of stamens in the normal and mutant plants. Am J Bot 60: 514-523
21. SAWHNEY VK, RI GREYSON 1973 Morphogenesis of the stamenless-2 mutant in tomato. II. Modifications of sex organs in the mutant and normal flowers by plant hormones. Can J Bot 51: 2473-2479
22. SAWHNEY VK, EB NAVE 1986 Enzymatic changes in post-meiotic anther development in *Petunia hybrida*. II. Histochemical localization of esterase, peroxidase, malate- and alcohol-dehydrogenase. J Plant Physiol 125: 467-473
23. VITHANAGE HIMV 1979 Pollen development and quantitative cytochemistry of exine and intine enzymes in sunflower, *Helianthus annuus*. Ann Bot (London) 44: 95-106
24. VITHANAGE HIMV, RB KNOX 1976 Pollen-wall proteins: quantitative cytochemistry of the origins of intine and exine systems in *Brassica oleracea*. J Cell Sci 21: 423-435

INSIGHTS INTO THE TEXAS MALE-STERILE CYTOPLASM OF MAIZE

Charles S. Levings, III and Carl J. Braun

Department of Genetics, Box 7614, North Carolina State University, Raleigh, NC 27695-7614, USA

Cytoplasmic male sterility (cms) is common among higher plants where it is reported in over 140 species (13). The trait is characterized by failure of viable pollen production and uniparental inheritance through the female (egg) parent (10). The non-Mendelian inheritance pattern suggests that cytoplasmic factors are responsible for cms, and indeed, recent studies, especially in maize and petunia (3, 26), indicate that mitochondrial gene mutations are the cause. Mitochondrial and chloroplast genomes are both generally transmitted only through the egg. It has been suggested that cms could be due to chloroplast genes, but there is little evidence in support. In a few cases, viruses may be responsible for cms. Cms has been used in hybrid seed production to eliminate the need for mechanical or hand emasculation, which are both expensive and tedious. Otherwise, cms has no apparent agronomic value.

Cms is manifested in a variety of ways among and within various plant species (10). In certain tobacco cms, sterility is associated with abnormal anther development (9). For example, anthers develop into a stigma-like structure (feminization) in the [rep] cms, while in the [und] cms the anther develops petals (petalody). In contrast, maize cms are very different from these tobacco cms in their expression. In *cms-T* maize, anthers are formed but pollen formation is terminated during microsporogenesis (23). In another maize cms, the *cms-S*, pollen abortion occurs late when nearly mature pollen grains suddenly abort (16). It is evident that cms can be caused by mechanisms affecting different developmental stages.

Cms is almost certainly caused by a variety of mutant genes. This is indicated by the wide assortment of mitochondrial gene mutations that are implicated with cms. In *cms-T*, a unique gene (designated *T-urf13* [*urf13*]) is believed to be responsible for cms (3). *urf13* is absent from normal and other male-sterile maize cytoplasms and, indeed, from other plant species. A mitochondrial gene, *Pcf*, is thought to cause cms in petunia (26). The *Pcf* gene is unique to the petunia cms and like *urf13* in maize, is a chimeric gene that arose by recombinational events involving other mitochondrial genes. Gene mutations are commonly due to mtDNA rearrangements in plants (17). Different mitochondrial gene mutations are also associated with the cms trait in other species. Although it is not unequivocally established that they cause cms, current evidence strongly implicates these genes with cms. It is this collective evidence which indicates that cms is due to mutations affecting different mitochondrial genes.

cms-T

Among cms, the Texas or *cms-T* cytoplasm of maize has received the most attention. *Cms-T* is distinguished from other maize cms by two nuclear restorer genes, *Rf1* and *Rf2*, that restore *cms-T* plants to pollen fertility (14). Both dominant alleles, *Rf1* and *Rf2*, are needed to suppress pollen sterility. There are two other major cms of maize, *cms-C* and *cms-S*. These cms are restored to pollen fertility by the *Rf4* and *Rf3* genes, respectively, but they are not restored to fertility by *Rf1* and *Rf2*. The specificity of restorer genes provides a means of differentiating the three cms and indicates that these cms are caused by different factors.

Cms-T is marked by failure of anther exertion and by pollen abortion. Like most other cms, female fertility is unaffected in *cms-T* plants. Uniquely associated with *cms-T* is susceptibility to the fungal pathogen, *Bipolaris maydis*, race T, (Southern corn leaf blight); other maize cytoplasms including *cms-C* and *cms-S* are resistant to race T (15). Prior to 1970, *cms-T* was extensively used to produce hybrid maize seed. In 1969 and 1970, an outbreak of Southern corn leaf blight forced the seed industry to curtail use of the Texas cytoplasm in hybrids (21).

The mitochondrial genome of *cms-T* codes for a gene, designated *urf13*, that is not encountered in other mitochondrial genomes (3). The gene appears to have originated by intramolecular recombinational events, because *urf13* consists partly of sequences with significant homology to other mitochondrial genes. The unusual origin undoubtedly accounts for the fact that *urf13* is limited to *cms-T*.

A 13-kD polypeptide is encoded by *urf13* that is expressed in all organs and tissues of *cms-T* and is localized in the inner mitochondrial membrane (5, 24). Since the *urf13* polypeptide is unknown in other plant mitochondria, it is presumed to be a novel and functionally unnecessary component of the membrane. The *Rf1* nuclear gene, which with *Rf2* restores pollen fertility to *cms-T* plants, influences the expression of *urf13* (5). Abundance of the 13-kD polypeptide is reduced about 80% in *cms-T* plants carrying the dominant allele *Rf1*. *Rf1* appears to regulate expression of *urf13* at the transcriptional or post-transcriptional level, because the transcriptional pattern of *urf13* is altered by the gene. *Rf2* does not affect expression of *urf13*, and it is unclear how it contributes to the restoration of pollen fertility.

DISEASE SUSCEPTIBILITY

The fungal pathogen *Bipolaris maydis*, race T, produces a pathotoxin, called T-toxin, that adversely affects mitochondria from *cms-T* plants, but does not effect mitochondria from other maize cytoplasms or other plant species. The toxin causes mitochondrial swelling, massive ion leakage from mitochondria, inhibition of respiration, and uncoupling of oxidative phosphorylation (1, 11, 18, 19). When calli initiated from embryonic tissue of *cms-T* are grown on medium containing T-toxin, toxin-resistant calli can be selected. Regeneration from toxin insensitive calli results in many regenerant plants that are pollen-fertile and resistant to *B. maydis*, race T, and its toxin (8). Fertile and resistant plants have also been obtained from similar experiments when toxin selection is omitted (2). Revertants from these studies are always both pollen-fertile and disease-resistant, indicating that the two traits are inseparable. They could be inseparable because the two traits are coded by a single gene or by two closely linked genes. Molecular investigations of several revertants have shown that *urf13* is deleted, probably by homologous recombination (20). In addition, an analysis of the unique revertant, T-4, has revealed a frameshift mutation in the *urf13* sequence that introduces a premature stop codon (25). Because mutations affecting the *urf13* gene are positively associated with the loss of cms and disease susceptibility, *urf13* is believed responsible for both traits.

Toxin sensitivity in *cms-T* mitochondria is due to the interaction of toxin with the 13-kD polypeptide encoded by *urf13* (4). This has been convincingly established in *E. coli* experiments where the *urf13* mitochondrial gene is transformed into bacterial cells and expressed. Cells expressing the URF13 protein are sensitive to T-toxin and methomyl, but cells not expressing the protein are insensitive to these compounds. Methomyl is an insecticide that affects *cms-T* mitochondria in the same fashion as T-toxin (12). Importantly, the *urf13* gene alone is able to confer toxin sensitivity to *E. coli* cells. Toxin effects in *E. coli* are analogous to the effects observed in *cms-T* mitochondria; they include the inhibition of bacterial respiration and growth, massive ion leakage and spheroplast swelling (4, C. Braun, unpublished results). The interaction of toxin and the 13-kD polypeptide causes sensitivity by permeabilizing the membranes and allowing ion leakage. These results furnish compelling evidence that the URF13 polypeptide is responsible for the specific toxin sensitivity and disease susceptibility characteristic of *cms-T* maize.

Site-directed mutational experiments show that the *urf13* gene can be altered so that it is unable to confer toxin sensitivity to *E. coli*. Several different amino acid substitutions and deletions in URF13 abolish the polypeptide's capacity to cause toxin sensitivity and indicate that *urf13* could be changed so that it does not cause disease susceptibility in *cms-T* maize. Toxin-binding studies have recently revealed that T-toxin binds to the URF13 polypeptide (C. Braun, unpublished results). Moreover, analyses of toxin-insensitive *urf13* mutants show that toxin binding is reduced. These findings suggest that toxin binding is involved in the toxin-URF13 interaction. Continuation of these studies is providing additional insight into the mechanism of toxin binding and sensitivity. Finally, we are especially interested in determining whether mutations that abolish toxin sensitivity also eliminate cms.

MECHANISM FOR *cms-T*

Although substantial progress has been made in understanding how the URF13 protein is implicated in disease susceptibility, little is known about its role with cms. The *urf13* gene and its product are convincingly associated with the cms trait by molecular analysis of revertants and its interaction with the restorer gene. There are no conclusive data suggesting a mechanism by which the *urf13* gene product is able to cause pollen abortion. In fact, it is puzzling that a mitochondrial gene mutation would specifically interfere with pollen formation and yet apparently not affect other plant developmental processes. Plants carrying T cytoplasm are vigorous and productive, as evidenced by their extensive use in hybrid maize production prior to the Southern corn leaf blight epidemic in 1969 and 1970. Furthermore, cms is generally not detrimental to crop production in other species as well (*e.g.* sorghum). These findings suggest that pollen formation has a unique dependence on mitochondrial function that is not characteristic of other plant developmental processes.

A proposal explaining why mitochondrial gene mutations cause cms can be formulated based upon the results of prior studies. The proposal suggests that mitochondrial gene mutations associated with cms slightly impair mitochondrial function but, in most plant developmental processes, impairment is too slight to be limiting. In a more demanding process, a small amount of mitochondrial dysfunction could be limiting so that it disturbs the process. If pollen development demands high levels of mitochondrial function, then mitochondrial dysfunction associated with mutant gene products could lead to pollen abortion. During microsporogenesis, a 20- to 40-fold increase in mitochondrial number is observed in the tapetal cell layer of the anther and in developing microspores of maize (23). This amplification clearly indicates an extraordinary involvement of mitochondria in pollen formation. A precocious degeneration of tapetal cells is reported to be associated with the onset of pollen abortion in *cms-T* (23). Although a relationship between mitochondrial activity and cell degeneration is unestablished, it is possible that mitochondrial dysfunction is involved.

Mitochondrial genes encode polypeptides that are largely localized in the inner mitochondrial membrane and are components of the electron transport chain or the F_1-F_0 ATPase (6). These polypeptides are crucial for proper mitochondrial respiration and mutant forms of these polypeptides could cause respiratory deficiencies. A recent investigation in humans has demonstrated that a mitochondrial gene mutation can exhibit its deleterious effect in specific tissues. Leber's hereditary optic neuropathy causes optic nerve degeneration and cardiac dysrhythmia (22). This trait has been traced to a mitochondrial gene mutation resulting in a single amino acid change (arg → his) in the NAD-4 polypeptide. The polypeptide is a component of Complex I of the electron transport chain. It has not been determined why this mutant gene product specifically affects the optic nerve or heart function. This result is important because it establishes that mitochondrial gene mutations may not produce deleterious effects in all tissues despite the fact that the altered gene product is expressed in all cell types. Cms expression could be analogous in that the mitochondrial gene mutation may only express its deleterious effect in tissues involved in pollen formation while other tissues are unaffected.

Plant mitochondrial genes code for tRNAs and ribosomal RNAs, as well as at least one ribosomal protein, that are components of the mitochondrial protein synthesizing apparatus (6). Mutant genes for these products could also be responsible for cms because protein synthesis is necessary for mitochondrial biogenesis.

There is an alternative explanation for cms that is unique to *cms-T*. Flavell has suggested that an anther-specific substance exists which affects mitochondria in a manner similar to the T-toxin of *Bipolaris maydis*, race T (7). This proposal was made before the *urf13* gene and its product were discovered. Based upon our current understanding, the anther-specific substance could interact with the URF13 polypeptide to inhibit mitochondrial respiration and uncouple oxidative phosphorylation. If the resultant loss of mitochondrial function occurs in tissues or cells involved in pollen formation (*e.g.* tapetum), viable pollen would not be produced. Since the URF13 polypeptide is uniquely expressed in *cms-T* mitochondria, the anther-specific substance would not effect pollen formation in other maize cytoplasms. Thus far, an anther-specific substance with these properties has not been reported; nevertheless, it is an interesting proposal in light of recent discoveries regarding the interaction of toxin with the URF13 polypeptide.

Acknowledgments--We wish to acknowledge support provided by the National Science Foundation and Agrigenetics Research Corporation.

LITERATURE CITED

1. BEDNARSKI MA, S IZAWA, RP SCHEFFER 1977 Reversible effects of toxin from *H. maydis* race T on oxidative phosphorylation by mitochondria of maize. Plant Physiol 59: 540-545
2. BRETTELL RIS, E THOMAS, DS INGRAM 1980 Reversion of Texas male-sterile cytoplasm maize in culture to give fertile, T-toxin resistant plants. Theor Appl Genet 58: 55-58

3. DEWEY RE, CS LEVINGS III, DH TIMOTHY 1986 Novel recombinations in the maize mitochondrial genome produce a unique transcriptional unit in the Texas male-sterile cytoplasm. Cell 44: 439-449
4. DEWEY RE, JN SIEDOW, DH TIMOTHY, CS LEVINGS III 1988 A 13-kilodalton maize mitochondrial protein in *E. coli* confers sensitivity to *Bipolaris maydis* toxin. Science 239: 293-295
5. DEWEY RE, DH TIMOTHY, CS LEVINGS III 1987 A mitochondrial protein associated with cytoplasmic male sterility in the T cytoplasm of maize. Proc Natl Acad Sci USA 84: 5374-5378
6. ECKENRODE VK, CS LEVINGS III 1986 Maize mitochondrial genes. In Vitro Cell Dev Biol 22: 169-176
7. FLAVELL R 1974 A model for the mechanism of cytoplasmic male sterility in plants, with special reference to maize. Plant Sci Lett 3: 259-263
8. GENGENBACH BG, CE GREEN, CM DONOVAN 1977 Inheritance of selected pathotoxin resistance in maize plants regenerated from cell cultures. Proc Natl Acad Sci USA 74: 5113-5117
9. GERSTEL DU 1980 Cytoplasmic male sterility in *Nicotiana*. Tech Bull No 263 NCARS, pp 1-31
10. HANSON MR, MF CONDE 1985 Functioning and variations of cytoplasmic genomes: lessons from cytoplasmic-nuclear interactions affecting male fertility in plants. Int Rev Cytol 94: 213-267
11. HOLDEN MJ, H SZE 1984 *Helminthosporium maydis* T toxin increased membrane permeability to Ca^{2+} in susceptible corn mitochondria. Plant Physiol 75: 235-237
12. KOEPPE DE, JK COX, CP MALONE 1978 Mitochondrial heredity: a determinant in the toxic response of maize to the insecticide methomyl. Science 201: 1227-1229
13. LASER KD, NR LERSTEN 1972 Anatomy and cytology of microsporogenesis in cytoplasmic male sterile angiosperms. Bot Rev 38: 425-454
14. LAUGHNAN JR, S GABAY-LAUGHNAN 1983 Cytoplasmic male sterility in maize. Annu Rev Genet 17: 27-48
15. LEAVER CJ, MW GRAY 1982 Mitochondrial genome organization and expression in higher plants. Annu Rev Plant Physiol 33: 373-402
16. LEE S-LJ, ED EARLE, VE GRACEN 1980 The cytology of pollen abortion in S cytoplasmic male-sterile corn anthers. Am J Bot 67: 237-245
17. LEVINGS III CS, GG BROWN 1989 Molecular biology of plant mitochondria. Cell (in press)
18. MATTHEWS DE, P GREGORY, VE GRACEN 1979 *Helminthosporium maydis* race T toxin induces leakage of NAD+ from T cytoplasm corn mitochondria. Plant Physiol 63: 1149-1153
19. MILLER RJ, DE KOEPPE 1971 Southern corn leaf blight: susceptible and resistant mitochondria. Science 173: 67-69
20. ROTTMAN WH, T BREARS, TP HODGE, DM LONSDALE 1987 A mitochondrial gene is lost via homologous recombination during reversion of CMS T maize to fertility. EMBO J 6: 1541-1546
21. ULLSTRUP AJ 1972 The impacts of the Southern corn leaf blight of 1970-1971. Ann Phytopathol 10: 37-50

22. WALLACE DC, G SINGH, MT LOTT, JA HODGE, TG SSHURR, AMS LEZZA, LJ ELSAS II, EK NIKOSKELAINEN 1989 Mitochondrial DNA mutation associated with Leber's hereditary optic neuropathy. Science 242: 1427-1430
23. WARMKE HE, S-LJ LEE 1977 Mitochondrial degeneration in Texas cytoplasmic male-sterile corn anthers. J Heredity 68: 213-222
24. WISE RP, AE FLISS, DR PRING, BG GENGENBACH 1987 *urf13*-T of T cytoplasm maize mitochondria encodes a 13 kD polypeptide. Plant Mol Biol 9: 121-126
25. WISE RP, DR PRING, BG GENGENBACH 1987 Mutation to male fertility and toxin insensitivity in Texas (T)-cytoplasm maize is associated with a frameshift in a mitochondrial open reading frame. Proc Natl Acad Sci USA 84: 2858-2862
26. YOUNG EG, MR HANSON 1987 A fused mitochondrial gene associated with cytoplasmic male sterility is developmentally regulated. Cell 50: 41-49

ANTHER-SPECIFIC GENES: MOLECULAR CHARACTERIZATION AND PROMOTER ANALYSIS IN TRANSGENIC PLANTS

SHEILA MCCORMICK, DAVID TWELL, ROD WING, VIRGINIA URSIN, JUDY YAMAGUCHI, AND SUSAN LARABELL

Plant Gene Expression Center, USDA-ARS, Albany, CA 94710, USA

The study of pollen development offers access to many of the central questions of developmental biology, including differential gene expression and cell-to-cell interactions. We are particularly interested in using pollen expressed promoters to determine cis- and trans-acting factors that regulate specific gene expression in gametophytic cells, and in assessing the role of pollen wall proteins in the interactions between pollen and pistil.

DEVELOPMENTAL EXPRESSION, LOCALIZATION

We previously reported the isolation and preliminary characterization of cDNAs that are specifically expressed in tomato anthers (11). We have now characterized five of these clones (pLAT51, pLAT52, pLAT56, pLAT58, and pLAT59) in greater detail. Developmental Northern blots show that these late anther tomato (LAT) clones are anther-specific, which we define as at least 200-fold higher expression in anthers than in other tissues (pLAT52 is exceptional, in that it also shows expression in petal, 20- to 50-fold less than in anther). These five clones are all maximally expressed in pollen. During anther development, some of these clones are expressed in immature anthers at the tetrad stage of microsporogenesis, while others are first expressed in green petal stage anthers, when the microspores are binucleate (Table I).

We have used (Ursin *et al.*, in preparation) *in situ* hybridizations (12) to further localize the expression of these clones within the anther (Table II). These experiments show that RNA corresponding to these clones is present in microspores and pollen, confirming results obtained by Northern blot analysis. Interestingly, RNA corresponding to these clones is also present in

Table I. *Developmental expression of cDNA clones isolated from a mature anther library.*

	Pollen	MA	GPA	IA
LAT51	+++	++	+	-
LAT52	++++	+++	++	+
LAT56	+++	++	+	-
LAT58	+++	++	+	+/-
LAT59	+++	++	+	-

[a]MA = mature anther, GPA = green petal anther, IA = immature anther.

Table II. *In situ localization of RNA corresponding to LAT cDNAs*

	Pollen Tube	MA		GPA	
		Pollen	Wall	Microsp.	Wall
LAT51	+	+++	+++	++	+
LAT52	+	+++	+++	++	+
LAT56	+	++	++	+	-
LAT58	+	+++	+++	++	+
LAT59	+	++	++	-	-

the anther wall, suggesting that the clones are both gametophytically and sporophytically expressed. *In situ* hybridizations to several genetically male sterile anthers show no hybridization signal in the anther wall, and Northern blots of male sterile anthers show greatly reduced or no RNA corresponding to the LAT clones. RNA corresponding to these clones is still present in pollen that has germinated *in vitro* for 19 h. Taken together, these facts suggest that the clones we are studying show a post-meiotic expression pattern, and further suggest that the transcripts code for proteins that are required during pollen maturation, germination, or fertilization.

PROMOTER ANALYSIS

Genes that show the same tissue specificity and/or timing of expression may be homologous in the 5' noncoding regions, where factors that determine specific transcription bind. To identify such regions, we isolated several anther-specific promoters in order to compare them. We isolated genomic clones for three of the anther-specific cDNAs (LAT52, LAT56, and LAT59). Computer homology searches with the three upstream (promoter regions) of these genomic clones show no obvious similarities. We have determined that 1.4 Kb of 5' flanking DNA of LAT59 is sufficient to direct pollen expression of reporter genes (see below), so it is not obvious why no similarities are detected between these promoters. Deletion analysis and gel retardation experiments may reveal short, or somewhat degenerate, regions of homology that were not noticed in the computer alignments.

A chimeric transcriptional gene fusion of 1.4 Kb of 5' flanking DNA of the LAT59 gene fused to coding DNA of the bacterial -glucuronidase (GUS) reporter gene (5) was constructed and introduced into tobacco and tomato by *Agrobacterium*-mediated transformation (10). The resulting transgenic plants were assayed for GUS enzyme activity; the patterns of activity were similar for tomato and tobacco. Figure 1 (top) shows the GUS enzyme activity for a representative tomato plant containing the chimeric gene. Expression in mature anthers was 26-fold higher than in roots and 200- to 700-fold higher than in other organs. Results in Figure 1 (bottom) show that GUS activity is first detectable in green petal stage anthers. This result mimics the Northern blot analysis of "native" LAT59 mRNA. Independent transformants show the expected quantitative variation in the level of GUS activity in anthers; the timing of appearance of GUS activity during anther development is invariant, always at the green petal stage.

GUS activity in sporophytic tissues of the anther was assayed separately from activity in pollen by washing pollen out of the locules, and assaying the washed out anther walls and pollen separately. GUS activity was approximately 100-fold higher in pollen than in washed anther walls (Fig. 2, top). This demonstrates that the vast majority of GUS activity in anthers is due to enzyme in the pollen, rather than the sporophytic tissue of the anther. This raises the question of whether the RNA detected in anther walls by *in situ* analysis is translated, and perhaps whether the 1.4 Kb promoter region of LAT59 is sufficient to direct expression in the anther wall. We are currently investigating these questions.

In addition to the low but significant GUS expression in roots, we have also detected similarly low levels of expression in the R_1 seed (both embryo and endosperm) of the R_0 plants. We have been unable to detect an RNA signal corresponding to LAT59 in roots or seeds of untransformed plants. Either these low levels of reporter gene expression reflect extremely low levels of LAT59 expression in these tissues, or the 1.4 Kb region of the LAT59 promoter allows a leakier expression than the "complete" LAT59 promoter. If LAT59 is indeed expressed both in pollen and in the next generation seed, it represents a member of the "sporophytic/gametophytic overlap" genes (15).

These experiments indicate a specificity for the 1.4 Kb LAT59 promoter suitable for expressing genes that may specifically disrupt pollen development or function. Our initial experiments to construct a "male sterile" gene use the auxin and cytokinin genes from *Agrobacterium tumefaciens*. We have constructed chimeric genes that involve the fusion of the LAT59 promoter to the *tms2* gene (3) and to the *tmr* gene (1). Transgenic tomato plants containing the LAT59-*tmr* construct appear normal, suggesting that the regulation of the

LAT59 promoter is sufficient to avoid any cytokinin effects in the vegetative portions of the plant. We are currently examining the pollen fertility, pollen germination, and fruit set on these plants.

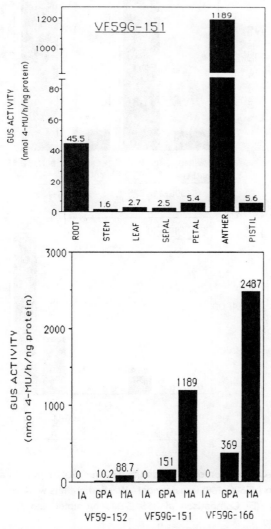

Fig. 1. (A) GUS activity in tissues of a tomato plant transformed with the LAT59(1.4)-GUS construct. (B) Developmental pattern of anther GUS activity in independent tomato transformants.

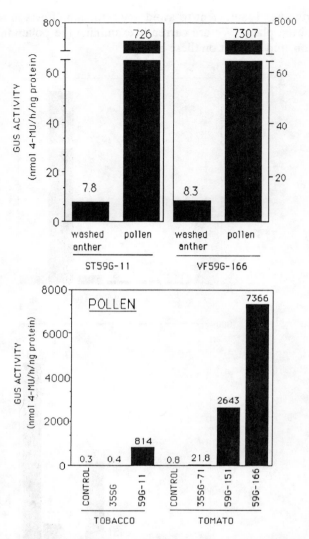

FIG. 2: (A) Comparison of pollen and anther wall GUS activity in a tobacco (ST59G-11) and a tomato (VF59G-166) transformant. Washed anthers were washed free of pollen before assaying. (B) Comparison of 35S CaMV promoter and LAT59 (1.4) promoter activities in pollen.

PROTEIN HOMOLOGIES

One reason we selected these five cDNAs for further study was that they did not cross-hybridize by Southern blot analysis, and were therefore assumed to correspond to different anther-specific genes. Surprisingly LAT56 and LAT59 were homologous at both the DNA (61%) and amino acid (54%) levels. Furthermore, these two genes are closely linked (approximately 5 map units apart), suggesting that they are products of a gene duplication event.

Computer searches with the open reading frames of LAT56 and LAT59 identified homologies to the pectate lyase genes of the bacterial plant pathogen *Erwinia* (Wing et al., in preparation). The pectate lyase genes are of interest in plant pathology because they are thought to be major determinants of the virulence of the *Erwinia* strains that cause soft rot disease (6). In plants, pectin degradation plays a role in fruit ripening, cell elongation and cell recognition (4). In pollen, pectin degrading enzymes may be used during pollen tube growth, in order to provide materials needed for wall synthesis (2).

The numerous pectate lyase genes of *Erwinia* show no sequence conservation except in two regions (I and II) of 18 and 22 amino acids (6, 8, 9, 14). LAT56 and LAT59 show 58% amino acid sequence identity to at least one of the pectate lyase genes in region I and region II, and additionally, are 51% homologous in the 33 amino acids between region I and region II, and 45% homologous in a 29 amino acid region located 5' to region I. A summary of the homologies in the area spanning region I to region II are shown in Table III. We are currently investigating whether the LAT56 and LAT59 proteins bind pectin or have pectate lyase activity.

Currently only a short region of open reading frame for pLAT58 has been identified, but pLAT52 and pLAT51 are full-length cDNAs. pLAT51 codes for a putative protein of 555 amino acids, and pLAT52 (Twell et al., in press) for a protein of 18 kD. Computer searches with these protein sequences have not revealed homology with any previously sequenced proteins.

Table III. *Percent of amino acid identities between LAT56, LAT59 and the pectate lyases of Erwinia chrysanthemi (Ch), and E. carotovora (Ca) in the amino acids spanning region I through region II*

	59	56	ACh	ECh	BCh	CCh	ACa	BCa
LAT59		83	53	60	50	60	38	48
LAT56			56	57	49	56	53	52
ACh				80	51	58	58	59
ECh					63	57	59	59
BCh						83	77	80
CCh							77	81
ACa								92

Pollen is known to release proteins from the wall within seconds after hydration (7); such proteins are presumably secreted. We suspect that the proteins encoded by LAT51, LAT52, LAT56, and LAT59 may be secreted, because all four proteins possess hydrophobic N-terminal regions with characteristics of signal sequences (16), and they each have at least one predicted N-linked glycosylation site. We are raising antibodies to the proteins to determine more precisely their cellular or extracellular localization.

POLLEN DEVELOPMENT AND FUNCTION

We want to know if the expression of any of these five genes is critical for pollen development or function. We are using an approach based on the expression of antisense RNA (13), with the hope that a decrease in expression of the LAT genes will be associated with an observable or measurable phenotype. Antisense constructs were made with the LAT51, LAT52 and LAT56 coding regions, initially using the 35S-promoter from Cauliflower mosaic virus (CaMV). Antisense RNA was expressed in leaf and anther tissue of transformed plants containing these constructs. We have not yet seen any altered phenotypes in these plants. One problem may be that the 35S CaMV promoter has very low activity in pollen, as compared to the LAT59 (1.4) promoter (Fig. 2, bottom). Currently, new antisense constructs are being made using our pollen-specific promoters, in order to maximize antisense expression in pollen.

The LAT clones are expressed in several other species of *Lycopersicon*, and in tobacco and petunia. The clones hybridize to homologous sequences in genomic DNA of *Arabidopsis* and corn. We hope to use these clones as molecular markers to study pollen development in other plants. One question that interests us is whether the LAT genes are transcribed from the vegetative or generative nucleus. We can answer that question more easily in corn, where we can obtain massive quantities of pollen.

We have shown that a 1.4 kb region of upstream sequences from LAT59 can direct pollen expression of reporter genes, in both tomato and tobacco (Twell *et al.*, in preparation). Further work will be required to determine if the low levels of expression we see in root and seed are due to naturally occurring low-level expression of LAT59 in these tissues, or due to the nature of the chimeric gene fusions of the promoter region that we used.

We plan to use the anther-specific promoters to express toxic genes, in order to construct artificial male steriles. In addition, we are continuing to delimit the DNA sequences necessary for the anther specificity of these promoters. Our future work on the coding regions of the anther-specfic genes will focus on protein secretion and processing.

LITERATURE CITED

1. Buchmann I, F-J Marner, G Schroder, S Waffenschmidt, J Schroeder 1985 Tumour genes in plants: T-DNA encoded cytokinin biosynthesis. EMBO J. 4: 853-859
2. Heslop-Harrison J 1987 Pollen germination and pollen-tube growth. In KL Giles, J Prakash, eds, Pollen: Cytology and Development, International Review of Cytology vol. 107. Academic Press, New York, pp 1-78
3. Inze D, A Follin, M Van Lijsebettens, C Simoens, C Gentetello, M Van Montagu, J Schell 1984 Genetic analysis of the individual T-DNA genes of Agrobacterium tumefaciens; further evidence that two genes are involved in indole-3-acetic acid synthesis. Mol Gen Genet 194: 265-274
4. Jarvis MC 1984 Structure and properties of pectin gels in plant cell walls. Plant Cell Environ 7: 153-164
5. Jefferson RA, TA Kavanagh, MW Bevan 1987 GUS fusions: beta-glucuronidase as a sensitive and versatile gene fusion marker in higher plants. EMBO J 6: 3901-3907
6. Keen NT, S Tamaki 1986 Structure of two pectate lyase genes from Erwinia chrysanthemi EC16 and their high level expression in Escherichia coli. J Bacteriol 168: 595-606
7. Knox RB 1984 Pollen-pistil interactions. In HF Linskens, J Heslop-Harrison, eds, Cellular Interactions, Encyclopedia of Plant Physiology, New Series, Vol 17. Springer-Verlag, Heidelberg, pp 508-608
8. Lei S-P, H-C Lin, S-S Wang, J Callaway, G Wilcox 1987 Characterization of the Erwinia carotovora pelB gene and its product pectate lyase. J Bacteriol 169: 4379-4383
9. Lei S-P, H-C Lin, S-S Wang, G Wilcox 1988 Characterization of the Erwinia carotovora pelA gene and its product pectate lyase A. Gene 62:159-164
10. McCormick S, J Niedermeyer, J Fry, A Barnason, R Horsch, R Fraley 1986 Leaf disc transformation of cultivated tomato (L. esculentum) using Agrobacterium tumefaciens. Plant Cell Rep 5: 81-84
11. McCormick S, A Smith, C Gasser, K Sachs, M Hinchee, R Horsch, R Fraley 1987 Identification of genes specifically expressed in reproductive organs of tomato. In D Nevins, R Jones, eds, Tomato Biotechnology. Alan R. Liss, New York, pp 255-265
12. Smith A, M Hinchee, R Horsch 1987 Cell and tissue specific expression localized by in situ RNA hybridization in floral tissues. Plant Mol Biol Rep 5: 237-241
13. Smith CJS, CF Watson, J Ray, CR Bird, PC Morris, W Schuch, D Grierson 1988 Antisense RNA inhibition of polygalacturonase gene expression in transgenic tomatoes. Nature 334: 724-726
14. Tamaki SJ, S Gold, M Robeson, S Manulis, NT Keen 1988 Structure and organization of the pel genes from Erwinia chrysanthemi EC16. J Bacteriol 170: 3468-3478
15. Tanksley SD, D Zamir, CM Rick 1981 Evidence for extensive overlap of sporophytic and gametophytic gene expression in Lycopersicon esculentum. Science 213: 453-455
16. Von Heijne G 1983 Patterns of amino acids near signal-sequence cleavage sites. Eur J Biochem 133: 17-21

MOLECULAR PHYSIOLOGY OF THE POLLEN STIGMA INTERACTION IN *BRASSICA*

CAROLE J. ELLEMAN, R. H. SARKER, G. AIVALAKIS, HELEN SLADE, AND H. G. DICKINSON

School of Plant Sciences, University of Reading, Whiteknights, Reading, RG6 2AS, UK (C.J.E., G.A., H.S., H.G.D.) and Department of Botany, University of Dhaka, Dhaka 1000, Bangladesh (R.H.S.)

Encouraging progress is currently being made with the identification and cloning of genes coding for stigmatic and stylar polypeptides potentially involved in self-incompatibility (SI) (2, 19), but a number of important questions remain to be answered concerning the more general nature of SI in flowering plants. Sporophytic and gametophytically controlled SI systems clearly differ in their structure and physiology, but we remain ignorant as to whether they share a common origin and, if so, at what stage divergence occurred. Secondly, although SI remains one of the few *bona fide* recognition events in higher plants, it is far from clear whether the signaling system involved follows the pattern now well characterized for animals (1), or represents an entirely new type of communication. In this connection, it might be profitable to enquire as to whether similar signaling systems are shared between SI and host-pathogen interactions (14). Most importantly, the SI response must be viewed in the perspective of the complex pollen stigma interaction of which it forms only part.

Brewbaker (3) was first to demonstrate that sporophytic and gametophytic control of pollen compatibility is accompanied by different morphological and physiological characteristics, and this work has been expanded by other authors (13). However, detailed examination of the cell biology of these events suggests that once compensation has been made for the fact that sporophytic systems are normally associated with dry stigmas, there may be fewer differences between the two processes than are currently supposed. For example, both appear to involve the participation of glycoproteins (GPs) with very high isoelectric points (pIs) (2, 18) which themselves are part of a large family of similar glycosylated polypeptides present in the pistil. Convincing evidence is also now available that, in both systems, pollen development appears to be arrested by a biostatic mechanism (17, 24).

Taken together, these data suggest that SI may represent a previously uncharacterized system of recognition and response. The presence of a GP family, which is also found in self-compatible varieties (23), indicates that evolution may have acted upon molecular species already involved in the pollen stigma interaction, and conferred a role in SI upon one or more members of this family. Because they may already be active in the pollen stigma interaction, it is possible that the SI role of GPs may not be as "true" signaling molecules (1), but rather in the modification of their established activity. The observation that self-pollen is held by a reversible biostatic mechanism supports such a view, since a 'conventional' signaling event would result in the rapid and total inhibition of male development.

The question of similarities between pollen stigma and host-pathogen interactions is considered extensively elsewhere (14, 15) but, in view of the very different philosophies of recognition implicit in the two interactions, too strong an analogy should not be drawn. Even were it accepted that the recognition process *per se* was different but the systems shared a response mechanism, this would again be unlikely since pathogenic challenge generally results in the synthesis of phytoalexin by the host, with the resulting necrosis of tissue adjacent to the pathogen. This would be a particularly foolish strategy for a stigma upon which compatible grains are also developing.

To appreciate the position of SI in the context of the pollination process *in toto*, it is necessary first to examine the nature of the compatible response, as it is seen in self-pollinating varieties, and then to determine how a system of SI could be overlaid upon it. While a number of principles developed in this work are equally applicable to sporophytic and gametophytic systems, we present here only data from recent work on the sporophytic SI system operating in *Brassica oleracea*.

THE COMPATIBLE POLLEN STIGMA INTERACTION IN *BRASSICA*

Following any compatible pollination in *Brassica*, the grain first adheres to the stigmatic papillae. While the pollen is carried to the stigma by an insect vector, the forces responsible for the initial binding are not yet clear. The grain coating, a tapetally-derived tryphine (6), is certainly sticky but there is persuasive evidence that the charge differential between pollen and stigma may be sufficient to generate electrostatic forces operating over distances of up to 1 mm (4, 5). Whatever mechanism controls the final approach of the grain to the stigma surface, contact is made initially between the superficial layer of the tryphine (8) and the stigmatic pellicle (16). The grain then extracts water from the stigma surface, drawing it through the coating into the protoplast. Recent experiments using anhydrous fixation techniques (8) indicate that the comparatively dry cytoplasm of the pollen undergoes spectacular reorganization as it rehydrates (Figs. 1 and 2). Equally dramatic is the change that overcomes the region of coating through which the water is passing into the pollen. Instead of a mottled, evenly staining matrix (Fig. 1), this area becomes strongly electron opaque, and its contents reorganized to form apparently membranous structures (8). The stigma itself also reacts to the presence of the pollen, and video microscopy shows dramatically that within a few minutes of pollination, cytoplasm aggregates under the point of contact (unpublished data). Further, alterations occur to the papillar wall beneath the grain including the apparent deposition of electron-opaque material immediately beneath the cuticle (Fig. 3), loosening of the fibrillar matrix (Figs. 4 and 7) and the formation of small globules within it (Fig. 4). After some 40 min, a tube is formed by the hydrated grain which grows first down into the area of coating between the grain and the stigma surface (Figs. 5-7), and then through the papillar pellicle into a specialized region between the cuticle and the normal cell wall (9).

The first result of this spectacular structural differentiation must be the establishment of hydraulic continuity between the male and female cells. This continuity is not established following interspecific pollinations outside the Brassicaceae, even if the other species possess dry stigmas (23). Some molecular congruity must, therefore, be necessary for the passage of water to occur; we do not yet know where these molecules are, but the fact that water

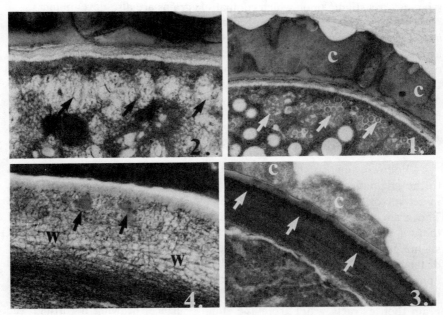

FIGS. 1-4. FIG. 1. "Dry" fixation of newly-released pollen showing homogenous coating (C) and spherical inclusions (arrows). x 8,500. FIG. 2. Partially-hydrated grain showing fibrillar layer (arrows) derived from spherical inclusions. x 16,800. FIG. 3. Electron-opaque deposition (arrows) in papillar wall beneath point of coating (C) contact. x 15,250. FIG. 4. Loosening of papillar wall (W) and presence of globules (arrows) beneath developing pollen grain. x 32,000.

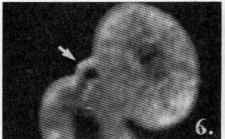
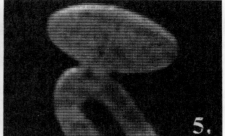

FIGS. 5 and 6. Pollen on stigmatic papilla 30 s and 30 min after pollination, demonstrating hydration of the grain and production of a pollen tube (arrow). [Photograph taken directly from video monitor.] x 850.

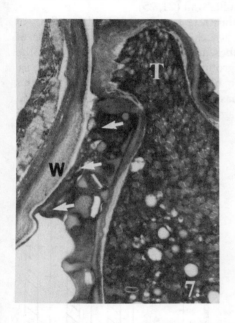

FIG. 7. Pollen developing on stigma surface showing tube (T) and loosening of fibrillar wall (W). Note transformed pollen coating (arrows). x 7,500.

will not flow between grains in contact with one another (24) suggests that some female participation is necessary. This is supported by the observation (26) that rinsing the stigma surface with water and then drying prevents hydration of compatible grains for at least 2 h, indicating not only that a superficial female molecule is involved, but also that it is capable of replacement.

Water is unlikely to be drawn into the pollen protoplast in isolation. Apart from the observation that *Brassica* pollen will not germinate in pure water, either supplied as a vapor or liquid, there is now persuasive evidence that molecules regulating hydration are also translocated to the pollen (24).

Interestingly, if bud stigmas are pollinated with mature pollen, the grains hydrate rapidly, often within 15 min, but if mature stigmas are pollinated with mature grains, hydration will take up to 60 min. A reasonable conclusion might be that structural changes during the development of the papillar wall restrict the passage of water, but treatment with cycloheximide (22, 24) produces hydration times very similar to those seen on buds (Fig. 8). Because of alleged nonspecific effects of cycloheximide, this work has been expanded to include the protein synthesis inhibitors puromycin and blasticydin S, with precisely similar results (Figs. 9 and 10). The conclusion is thus inescapable that a new compound is synthesized within the papillae which regulates the hydration of the pollen, and that inhibition of protein synthesis negates its effect. These surprising results underline the dynamic nature of the stigmatic papilla and its surface, and add to an accumulating body of evidence that there are a number of cycling polypeptides, some of them perhaps glycosylated, present in the papilla wall (21, 25).

The present challenge is to demonstrate the physiological functions of these molecules *in vitro*. Some time ago, it was shown that high levels of germination and growth of straight tubes in *Brassica* can be achieved if the medium contains high levels of amino groups, a "physical" molecule such as polyethylene glycol, and is buffered to high pH (20) (Fig. 11). While this work remains largely undeveloped, it has been discovered that stigmatic GPs, particularly those with high pIs, can dramatically promote the germination of

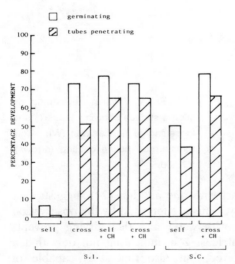

FIG. 8. Effect of cycloheximide (CH) on pollen germination and tube penetration following self- and cross-pollinations in self-incompatible (SI) and self-compatible (SC) systems.

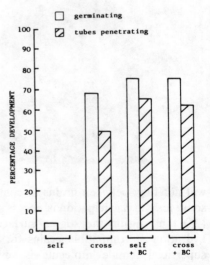

FIG. 9. Effect of blasticidin S (BC) on pollen germination and tube penetration following self- and cross-pollination in self-incompatible *Brassica oleraca*.

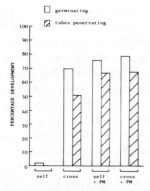

FIG. 10. Effect of puromycin (PM) on pollen germination and tube penetration following self- and cross-pollination in self-incompatible *Brassica oleraca*.

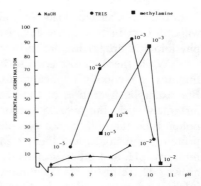

FIG. 11. The effect of percentage germination and pH of the medium after adding Tris, methylamine, and NaOH at different molarities. 500 grains were scored for each treatment.

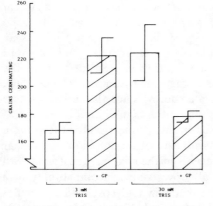

FIG. 12. The effect of adding stigmatic glycoprotein (GP) to germination media containing Tris at 3 and 30 mM. The GP fraction used was prepared by carboxymethyl sepharose chromatography followed by filter purification and covered a pI range from 8-10.5.

pollen and growth of tubes when added to the medium. By adjustment of the medium these GPs can be used to compensate for Tris, which presumably serves both as the buffer and to contribute the vital amino groups (Fig. 2). Bioassays of this type cannot claim to duplicate conditions *in vivo* where the system is essentially dry; for example, relatively small amounts of water must pass from the stigma to the grain *in vivo*, and stigmatic GPs, or other molecular species, are probably present at very high concentrations. We should therefore regard stigmatic molecules as not only possibly regulating pH, but also osmolarity and other essential parameters for pollen development. Unfortunately, the bioassay cannot be used to identify the stigmatic molecule which regulates pollen grain hydration.

Whether or not molecules from the sporophytically-derived coating of the pollen play a role in compatible development, other than in adhesion (25) and in the passage of water, remains unclear. The coating may be removed from the grain surface using organic solvents, and pollen treated in this way will still germinate very readily *in vitro*. The content of the coating, and its physiological properties, are further discussed in the next section.

While the data discussed so far relate solely to *Brassica oleracea*, many findings are equally applicable to gametophytically-controlled SI systems. For example, hydration of the grains on wet stigmas does not take place instantaneously, but over 15 min or so (17). Germination and growth of the pollen tube take place in a complex mixture of sugars, polypeptides, glycoproteins, and lipids, and there clearly must be elements which regulate the molarity, pH, and other properties of the medium through which the pollen grows. These pistils also contain a range of highly charged GPs resembling in some ways those described for *Brassica*, and which may fulfill very much the same functions.

SELF-INCOMPATIBILITY IN *BRASSICA*

The SI system must operate either within, or on, the processes described in the previous section. Although self-grains appear not to hydrate to the same degree as cross, some hydration always occurs (27) and a system which solely denies access of stigmatic water to the grain cannot thus be invoked. Equally, it is not practical to suggest that the stigma produces an inhibitor, or else compatible grains would also be inhibited. One or more molecular species with a cognitive function must thus be transferred from the stigma to the pollen grain during the first stages of hydration, whatever the compatibility of the pollination. There is some evidence that this molecule may be a cycling polypeptide, in that protein synthesis inhibitors overcome SI (24). Extreme caution has to be exercised in this connection for the protein synthesis inhibitors could be affecting synthesis of an inhibitor or similar molecule within the pollen, rather than events in the stigma. Treatment with the glycosylation inhibitor tunicamycin also overcomes incompatibility in a very spectacular fashion (24), suggesting that newly glycosylated molecules are required, rather than those from the large pools known to be present in the stigmatic protoplasts. A reliable bioassay would be a valuable tool in identifying these molecules but, although tube growth can be modified using stigmatic molecules as discussed earlier, *bona fide* reports of S-gene product activity *in vitro* have been few in species with sporophytic SI (10). Close examination of the situation *in vivo* has, however, indicated to us that a functional bioassay for *Brassica* S-gene products may be feasible, but using a radically different approach. This is very difficult work, and before the effect of potential SI effector molecules can be investigated, the promotive effects of individual stigmatic molecules must first be 'tuned out' of the system. We are now approaching this stage and, in our hands, the bioassay now demonstrates the inhibition of pollen tube growth by high pI stigmatic GPs derived from the same genotype. Treatment of pollen of different genotypes results in no such inhibition. The preliminary nature of this work means that we have yet to investigate whether the inhibition of 'self'-pollen is biostatic, as it clearly is *in vivo* (24). Central to the success of this bioassay has been the appreciation of the part played by the sporophytically-derived coating investing the pollen grain, for logic dictates that interaction between the female molecule and a coating-borne male protagonist results in the regulation of pollen development. We have now succeeded in isolating and characterizing the major polypeptides contained in the coating and, although this work will be the subject of a more extended report, it is noteworthy that the coat contains a family of highly charged GPs, ranging in pI between 8 and 10, and in mol wt between 20 and 60 kD. At present, our work suggests that none are identical with the so-called S-allele specific GPs of the female side, and we have yet to find convincing evidence of S-allelic differences between these polypeptides.

Despite the dramatic increase in molecular and physiological information available on SI in *Brassica*, only the vaguest picture is emerging as to how it may operate. Firstly, there is an accumulating body of evidence that the female protagonist is likely to be a GP, for which the gene may have already been cloned (19). The pedigrees of S-allele associated glycoproteins (18), developmental studies-particularly those involving cycloheximide (24, 21), the early data (10) on a putative SI product, and our bioassay using highly-charged GPs all support such an inference. However, if the active female molecule is a GP which represents a significant percentage of stigmatic total protein, it is unlikely that a conventional signaling reaction is involved. We have also discussed earlier that since this GP is one of a family contained in the stigma, it possibly possesses some other function in addition to its role in SI. In view of the high pI of this general class of molecule and the amine requirement for pollen germination, it is possible that it acts to stimulate pollen development very much as polyamines do elsewhere in the plant kingdom (11). Whether or not it regulates the development of self-pollen by modification of its 'normal' function or by a different route remains unclear.

At this level, parallels between gametophytic and sporophytic systems are obvious. *In vitro* experiments using a range of species, including *Lilium* (17), *Nicotiana* (2), *Papaver* (12) and *Antirrhinum* (A. McCubbin, unpublished), all point to high pI GPs regulating the development of pollen according to S-genotype. As in sporophytic systems, these GPs form part of the spectrum of complex molecules present in the pistil each presumably possessing a function in establishing the appropriate physico-chemical environment for development of the pollen tube. Even assuming this degree of commonality between sporophytic and gametophytic systems, we face only a slightly easier task in explaining the manner by which the two types of mechanisms may have evolved.

Acknowledgments--This work has been supported by the United Kingdom Agriculture and Food Research Council under its Cell Signalling Initiative.

LITERATURE CITED

1. ALBERTS B, D BRAY, J LEWIS, M RAFF, K ROBERTS, JD WATSON 1983 Molecular biology of the cell. Garland, New York, pp 718-763
2. ANDERSON MA, EC CORNISH, SL MAU, EG WILLIAMS, R HOGGART, A ATKINSON, I BONIG, B GREGO, R SIMPSON, PJ ROCHE, JD HALEY, JD PENSCHOW, HD NIALL, GW TREGEAR, JP COCHLAN, RJ CRAWFORD, AE CLARKE 1986 Cloning of a cDNA for a stylar glycoprotein associated with the expression of self-incompatibility in *Nicotiana alata*. Nature 321: 38-44
3. BREWBAKER JL 1957 Pollen cytology and incompatibility systems in plants. J Hered 48: 271-277
4. CHALONER WG 1986 Electrostatic forces and their significance in exine ornament. Pollen and Spores: Form and Function. Linn Soc Symp Ser 12: 103-108. Academic Press, New York
5. CORBET SA, J BEAMENT, D EISIKOWICH 1982 Are electrostatic forces involved in pollen transfer? Plant Cell Environ 5: 125-129
6. DICKINSON HG, D LEWIS 1973 The formation of the tryphine coating the pollen grains of *Raphanus* and its properties relating to the self-incompatibility system. Proc Roy Soc Lond B Biol Sci 184: 149-165
7. DICKINSON HG, JM MORIARTY, J LAWSON 1982 Pollen-pistil interaction in *Lilium longiflorum*: the role of the pistil in controlling pollen tube growth following cross- and self-pollinations. Proc Roy Soc Lond B Biol Sci 215: 45-62
8. ELLEMAN CJ, HG DICKINSON 1986 Pollen stigma interaction in *Brassica*. IV. Structural reorganisation in the pollen grain during hydration. J Cell Sci 80: 141-157
9. ELLEMAN CJ, CE WILLSON, RH SARKER, HG DICKINSON 1988 Interaction between the pollen tube and stigmatic cell wall following pollination in *Brassica*. New Phytol 109: 111-117
10. FERRARI TE, DH WALLACE 1975 Germination of *Brassica* pollen and expression of incompatibility *in vitro*. Euphytica 24: 757-765
11. FIENBERG AA, JH CHOI, WB LUBIGH, ZR SUNG 1984 Developmental regulation of polyamine metabolism in growth and differentiation of carrot culture. Planta 162: 532-539
12. FRANKLIN-TONG VE, MJ LAWRENCE, FCH FRANKLIN 1989 An *in vitro* bioassay for the stigmatic product of the self-incompatibility gene in *Papaver rhoeas*. New Phytol (in press)
13. HESLOP-HARRISON J, KR SHIVANNA 1977 The receptive surface of the angiosperm stigma. Ann Bot 4l: 1233-1258
14. HODGKIN T, HG DICKINSON, G LYON 1987 Self-incompatibility in *Brassica oleracea*: a recognition system with characteristics in common with plant-pathogen interactions? *In* GP Chapman, CC Ainsworth, CJ Chatham, eds, Eukaryotic Cell Recognition: Concepts and Model Systems. Cambridge Univ Press, Cambridge, pp 257-274

15. HODGKIN T, GD LYON, HG DICKINSON 1989 Recognition in flowering plants: a comparison of the *Brassica* self-incompatibility system and plant pathogen interactions. New Phytol (in press)
16. MATTSSON, O, RB KNOX, J HESLOP-HARRISON, Y HESLOP-HARRISON 1974 Protein pellicle of stigmatic papillae as a possible recognition site in incompatibility reactions. Nature 247: 298-300
17. MORIARTY JM 1984 Self-incompatibility in *Lilium longiflorum*. PhD Thesis, University of Reading, UK
18. NASRALLAH JB, RC DONEY, ME NASRALLAH 1985 Biosynthesis of glycoproteins involved in the pollen stigma interaction of incompatibility in developing flowers of *Brassica*. Planta 165: 100-107
19. NASRALLAH JB, SM YU, ME NASRALLAH 1988 Self-incompatibility genes of *Brassica oleracea*: expression, isolation and structure. Proc Natl Acad Sci USA 85: 5551-5555
20. ROBERTS IN, TC GAUDE, G HARROD, HG DICKINSON 1983 Pollen stigma interactions in *Brassica oleracea*: a new pollen germination medium and its use in elucidating the mechanism of self-incompatibility. Theor Appl Genet 65: 231-238
21. ROBERTS IN, G HARROD, HG DICKINSON 1984 Pollen stigma interactions in *Brassica oleracea* I. Ultrastructure and physiology of the stigmatic papillar cells. J Cell Sci 66: 241-253
22. ROBERTS IN, G. HARROD, HG DICKINSON 1984 Pollen stigma interactions in *Brassica oleracea* II. The fate of stigma surface proteins following pollination and their role in the self-incompatibility response. J Cell Sci 66: 255-264
23. SARKER RH 1988 Cytophysiology of self-incompatibility in *Brassica*. PhD Thesis, University of Reading, UK
24. SARKER RH, CJ ELLEMAN, HG DICKINSON 1988 Control of pollen hydration in *Brassica* required continued protein synthesis, and glycosylation is necessary for intraspecific incompatibility. Proc Natl Acad Sci USA 85: 4340-4344
25. STEAD AD, IN ROBERTS, HG DICKINSON 1979 Pollen pistil interactions in *Brassica oleracea*: Events prior to pollen germination. Planta 146: 211-216
26. ZUBERI MI, HG DICKINSON 1985 Pollen stigma interaction in *Brassica* III. Hydration of the pollen grains. J Cell Sci 76: 321-336
27. ZUBERI MI, HG DICKINSON 1985 Modification of the pollen stigma interaction in *Brassica oleracea* by water. Ann Bot 56: 443-452

THE GENETICS OF SELF-INCOMPATIBILITY REACTIONS IN *BRASSICA* AND THE EFFECTS OF SUPPRESSOR GENES

MIKHAIL E. NASRALLAH

Section of Plant Biology, Cornell University, Ithaca, NY 14853, USA

Self-incompatibility was observed in radish (*Raphanus sativus*) in a report dating back to 1920 (22). Microscopic examination of this species by Sears led him to the conclusion that incompatibility was due to the failure of pollen tubes to invade the stigma (21). However, the study of self-incompatibility began in earnest in the early 1950s with Bateman's analysis of the inheritance patterns in self-incompatibility in *Iberis*, and his interpretation of the results obtained by A G Brown at the John Innes Institute in *Raphanus* (2). Bateman explained the complex inheritance patterns by invoking the action of one locus with multiple alleles, sporophytic control of the pollen reaction, and the occurrence of codominant and dominant/recessive allelic interactions. The hypothesis of sporophytic control was verified by several workers in kale (24), radish (19), and several other cultivars of *Brassica oleracea* and *B. campestris*.

The number of alleles at the S-locus was estimated to be 22 in *Iberis* populations (2) and 25 to 34 in *R. raphanistrum* (20). Forty-one different alleles were enumerated in cultivated crops of *B. oleracea* (17), and the total number of alleles isolated to date in this species appears to have plateaued at 50 to 60 alleles.

Allelic interactions appear to vary with the species under study. In *R. raphanistrum*, Sampson (20) found dominance to be common in the pollen (18 out of 28), and rare in the stigma (4 out of 28) heterozygous combinations. In *R. sativus*, Tatebe (23) and Haruta (3) found dominance in the pollen, but only independent action in the stigma. The genus *Lesquerella* exhibited interactions similar to *Raphanus*, while in populations of *Iberis*, dominance was common in both pollen and stigma (1). Overall, it is estimated that 60% of allelic pairs exhibit codominance in crucifers. This is in contrast to the situation in species with gametophytic self-incompatibility, such as *Nicotiana*, in which codominance is the absolute rule. The occurrence of codominant interactions in gametophytic and sporophytic systems has favored a unifying model of plant self-incompatibility. In this model, recognition of self is proposed to be an "oppositional" system with positive inhibition arising from the interaction of identical alleles (10).

MOLECULAR MARKERS OF SELF-INCOMPATIBILITY

S-Locus Specific Glycoprotein (SLSG) Polymorphisms. The genetic analysis of self-incompatibility in *Brassica* was greatly aided by the identification in stigma extracts, first of S-specific antigens (15), and later of SLSG (4, 12, 16, 18). SLSG polymorphisms can most easily be detected on isoelectric focusing (IEF) gels, but have also been observed on SDS-PAGE (12). Genetic data from our laboratory, and from that of Hinata and colleagues, has demonstrated a perfect correlation of the segregation of these SLSGs and that of the corresponding S alleles. In view of this strict correspondence and of the fact that, at the protein level, codominant expression in the stigma of the SLSGs encoded by the two alleles seems to be the rule, the screening of a segregating population by SLSG polymorphism is very useful in assigning S-genotypes.

DNA Polymorphisms. SLSG patterns can only be identified at the flowering stage, and the time required for screening is therefore dependent on plant generation time. Recently, cDNA sequences encoding SLSG from *Brassica* (11), and the genes encoding them (SLG), were isolated (13). The use of these cDNA probes revealed extensive restriction fragment length polymorphisms (RFLPs) when genomic DNA from several different S-allele homozygous genotypes was analyzed by blot hybridization. Just like SLSG polymorphism, this DNA polymorphism has been shown to cosegregate with the corresponding S alleles and, in all cases analyzed to date, the S-locus genotype can be correctly inferred from the pattern of the homologous genomic restriction fragments. DNA markers provide for convenient and rapid analysis of segregating populations, and circumvent the time-consuming task of performing diallel pollinations. S-related DNA patterns can be deduced from genomic DNA samples prepared rapidly from a small amount of leaf tissue. Screening can therefore be accomplished very early at the seedling stage in order to establish putative incompatibility groups. Furthermore, in the occasional cases in which pollination analysis cannot distinguish between dominant homozygotes and heterozygotes because of poor pollen viability, for example, the use of protein and DNA markers allows for rapid and unambiguous genotype assignment.

Monoclonal Antibody Probes. The ability to identify S-gene products is also enhanced by the use of specific antibody probes. We have recently developed monoclonal antibodies (MAbs) directed against SLSG purified from *B. oleracea* S_6 homozygotes. When used in combination with an enzyme-linked detection system to stain blots of proteins resolved by SDS-PAGE, the sensitivity is such that SLSG from one stigma or less is easily detected. In this way, not only are the visualization of SLSG polymorphism and, consequently, the verification of S-genotype assigned by RFLP analysis greatly facilitated, but the analysis of variation in SLSG levels and investigation of SLSG function are also possible.

One monoclonal, MAbH8, recognizes SLSG via a protein epitope which has been mapped to the conserved amino-terminal domain of the molecule. MAbH8 reacts on SDS-PAGE with SLSGs from a wide variety of S-genotypes of *Brassica*, including *B. oleracea*, *B. campestris*, *B. napus*, and of *Raphanus sativus* and *R. raphanistrum*. In *B. oleracea*, out of 20 S-allele homozygotes tested to date, 16 reacted with MAbH8, and four others (the S_2, S_4, S_5, and S_{15} homozygotes) did not (8). In this latter group, S_2, S_5, and S_{15} were studied by Thompson (25) and found to be pollen-recessives. The molecular basis of the divergence of these alleles is currently under investigation in our laboratory.

INHERITANCE OF SELF-INCOMPATIBILITY IN *BRASSICA CAMPESTRIS*: AN EXAMPLE

In this paper, I report on a study of a *B. campestris* population derived from a self-incompatible heterozygous plant. The self-incompatibility phenotype of this "founder" plant was assayed by microscopic examination of pollinated stigmas as described below and verified by counting the seed set following self-pollination of mature and immature stigmas along a developing inflorescence. The heterozygosity of this plant was determined by electrophoretic analysis of stigma extracts. Stigma proteins were resolved by IEF, transferred to nitrocellulose by capillary blotting, and immunostained with MAbH8. Two reactive bands were visualized instead of the one band characteristic of most S-allele homozygous genotypes, and the plant was assigned the genotype S_2-S_3 (these S-allele numbers are arbitrary and do not imply identity with any other *B. campestris* S-alleles).

Genotype Characterization of Progeny Plants by RFLP Analysis. Approximately 100 seeds were obtained from the founder plant by selfing at the immature bud stage, and 30 progeny plants were analyzed at the DNA, protein, and phenotypic levels. DNA was prepared from a small amount of leaf tissue obtained at the seedling stage by a mini-preparation procedure modified from a rapid phage lysis protocol. Following electrophoresis in 0.9% (w/v) agarose of EcoRI digests, the DNA was transferred to Gene Screen Plus filters and probed with SLSG-encoding cDNA labeled with ^{32}P by random priming. As has been consistently observed in a number of *Brassica* strains (11, 13), the probe identified S-related polymorphisms and a number of different restriction fragments in each plant. Figure 1 shows that three different RFLP patterns can be identified. One pattern (pattern I), exhibited by plants #4, 6, 7, 10, and 11, is characterized by a prominant band in the 20 kilobases (kb) mol wt range and another at 1.4 kb. Another pattern (pattern III), exhibited by plants #2 and 5, is characterized by the presence of two prominant bands with mol wt > 10 kb, and the absence of the 1.4 kb fragment. Another pattern (pattern II), exhibited by plants #1, 3, 8, 9, 12, 13, 14, and 15, is a hybrid of the first two and is identical to that of the founder plant (not shown). Pattern II is therefore derived from the heterozygous plants, and patterns I and III can thus be assumed to be the two homozygous patterns. Pattern I has been arbitrarily designated as the S_2 homozygous pattern, and pattern III as the S_3 homozygous pattern. Among the 30 plants analyzed, eight exhibited RFLP pattern I, 15 exhibited RFLP pattern II, and seven exhibited RFLP pattern III (Table I). These numbers translate into a genotypic ratio for S_2S_2, S_2S_3, and S_3S_3 of 1:2:1.

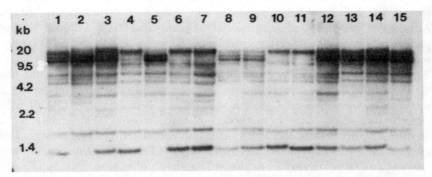

FIG. 1. RFLP analysis of a population of *B. campestris* plants segregating for two S-alleles. A blot of EcoRI-digested DNA was hybridized with an SLSG-cDNA probe. The numbers at the top of the figure are the numbers of the progeny plants. Mol wt markers are indicated.

Table 1. *Incompatibility Relationships of the Self-Incompatible Progeny Derived from One Heterozygous Self-Incompatible S2S3 Founder Plant*

Stigma Parent	Pollen Parent[a]		
	S_2S_2 Group I	S_2S_3 Group II	S_3S_3 Group III
S_2S_2 Group I	−	−	+
S_2S_3 Group II	−	−	−
S_3S_3 Group III	+	+	−
# Plants/Group	5	12	7

[a] + = compatible pollination; − = incompatible pollination

Genotypic Characterization of Progeny Plants by SLSG Polymorphism
After flowering, the 30 progeny plants were analyzed by protein immunoblotting. For each plant, the extract of one stigma was subjected to SDS-PAGE on 10% acrylamide slabs, electroblotted to nitrocellulose, and treated with MAbH8 followed by immunodetection with alkaline phosphatase-conjugated anti-mouse immunoglobulin (IgG). Here again, three patterns are evident (Fig. 2). One pattern, consisting of a cluster of three bands at 50 to 55 kD, is found in plants with RFLP pattern I (plants #6, 7, and 11), and is designated as the S2 homozygous pattern. Another pattern, consisting of a higher mol wt cluster, is found in plants with RFLP pattern III (plants #2 and 5), and is designated as the S_3 homozygous pattern. The third pattern is a composite of the first two, is exhibited by plants with RFLP pattern II (plants #1, 3, 8, 9, 12, 13, 14, and 15) and is the S_2S_3 heterozygous pattern. Among the 30 progeny plants analyzed, 24 plants had strong immunoreactive patterns. Based on a perfect correlation between RFLP and SLSG polymorphism, 5 were S_2S_2, 7 were S_3S_3, and 12 were S_2S_3.

Self-Incompatible Progeny. Phenotypic analysis of these 24 plants was carried out by one of two methods: the number of seed was counted after maturation, or pollen-tube growth was directly monitored by UV fluorescence microscopy. In this method, pollinated stigmas are fixed in a 3:1 mixture of ethyl alcohol and acetic acid, softened in 1N NaOH, and squashed after staining with decolorized aniline blue (9). Microscopic examination of a few stigmas (routinely 6) from the appropriate pollination, is usually sufficient to determine if the pollen reaction is compatible or incompatible. A series of diallel crosses among progenies of a heterozygous plant allows the definition of a finite number of incompatibility groups and the demonstration of the codominance or dominance-recessiveness of the pair of segregating S-alleles.

All 24 plants were shown to be like the founder plant, self-incompatible. The incompatibility relationships of these plants are shown in Table I. Based on their behavior in diallel crosses, they could be classified into three incompatibility groups. The plants within each group behaved identically with respect to plants from the other two groups. Pollen from group I and group II plants was incompatible with group I and group II stigmas, but compatible with group III stigmas; group III pollen was incompatible with stigmas from groups II and III, but compatible with group I stigmas. Based on DNA and protein polymorphisms, all plants in incompatibility group I were S_2S_2, those in group II were S_2S_3, and those in group III were S_3S_3. The identical responses of group I and group II plants when used as male parents points to the dominance of the S_2 allele over the S_3 allele in pollen. In the stigma however, S_2 and S_3 exhibit independent or codominant action, since group II stigmas are incompatible with pollen from all three groups.

Self-Compatible Progeny. The remaining six progeny plants had drastically reduced levels of S antigen, as shown for plants #3 and 10 in Figure 2. The genotype of these plants could still be determined on the basis

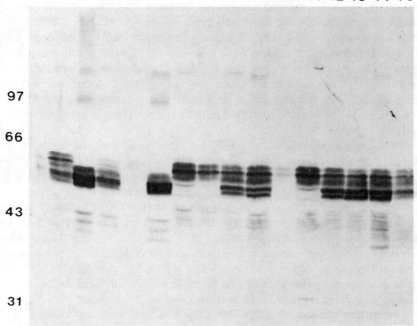

FIG. 2. Immunoblot analysis of a stigma extracts obtained from the plants shown in Fig. 1. The plant numbers at the top of the figure correspond to the numbers in Figure 1. Mol wt markers are indicated in kD to the left of the figure.

of SLSG polymorphism by overloading the lanes, and on the basis of their RFLP patterns: three were homozygous for the S_2 and three were S_2S_3 heterozygotes. Interestingly, these six plants were shown by phenotypic analysis to be self-compatible. Crosses to self-incompatible sibs have, in addition, shown that only the stigma reaction is altered in these self-incompatible plants, while the pollen reaction is unchanged.

The self-incompatible phenotype of the founder plant and the 3:1 ratio of self-incompatible to self-compatible plants in the progeny indicates that compatibility is due to the action of a single recessive suppressor gene which we will designate *sup2*. The occurrence of S_2S_2 and S_2S_3 self-compatible plants clearly indicates that the SUP2 gene lies outside the S-locus. Due to the fact that no S_3S_3 self-compatible plants were recovered, it can be assumed that the recessive *sup2* allele was in coupling with the S_2 allele in the founder plant, and that SUP2 and the S-locus are loosely linked to one another, and the linkage distance is currently being determined by recombination analysis.

GENE INTERACTIONS AND THE INHERITANCE OF SELF-COMPATIBILITY

The results of the genetic analysis reported here can be summarized as follows: (*i*) there is a perfect correlation between RFLP and SLSG protein pattern; (*ii*) plants with the same genotypic assignment based on RFLP and SLSG polymorphism fall into the same phenotypic incompatibility groups; (*iii*) a reduced level of stigma SLSG is correlated with a self-compatible stigma phenotype; (*iv*) self-compatibility in this population can be ascribed to the action of a recessive gene which lies outside the S-locus; and (*v*) the action of this gene is not specific to a particular S-allele since the level of SLSGs encoded by both S_2 and S_3 is equally reduced in self-compatible S_2S_3 plants.

The genetic breakdown of self-incompatibility in *Brassica* has been the subject of several reports. Although the generation of S_f (self-fertility) alleles at the S-locus has been discussed as a basis for self-compatibility, there is no evidence for the occurrence of such alleles in *Brassica*. Rather, it was shown, in cases where only the S-locus is involved, that self-compatibility results from the competitive interaction and mutual weakening of the two S-alleles (26). In the majority of cases, however, genes unrelated to the S-locus have been implicated. Self-compatibility, as described in lines of *B. oleracea* var *acephala* (26), in *B. oleracea* var *capitata* (14), and *B. campestris* (5, 6) was ascribed to single genes unlinked to the S-locus. In all three cases, loss of incompatibility was reported for the stigma, but not for pollen.

In studies of the self-compatible yellow sarson, a natural cultivar of *B. campestris* cultivated as an oil crop in India, Hinata and co-workers (5) have reported minimal or no activity of the S-locus, and have attributed self-compatibility to the action of an unlinked, recessive modifier gene designated "m". SLSG could still be detected in some self-compatible F_2 progeny of a cross between yellow sarson and a self-incompatible S_8 homozygote. The mode of action of the "m" gene is at present unknown. One suggestion proposed by the authors, is that it acts not at the level of pollen recognition, but at a subsequent step of pollen development.

In the case of *B. oleracea* var. *capitata* (14), self-compatibility was, on the other hand, shown to be correlated with a reduction of S-proteins in the stigma as demonstrated here for the suppressor gene 2 (*sup2*) mutant line. The suppressor gene (Su) implicated in self-compatibility in the *capitata* case was, however, semi-dominant in its action, unlike the gene described in this paper. In keeping with the convention of using three-letter gene designations, we will refer to the "m" genes as *mod1*, and the SU gene as SUP1.

From the examples discussed above, it is already clear that mutations in different modified genes act at different levels to functionally inactivate S-locus expression in the stigma (Table II). Although the mode of action has not been elucidated for any of the genes, I suggest that the breakdown of incompatibility occurs via at least two basic mechanisms.

Table 2. *Summary of Genetic Interactions*

Gene Symbol	Description	Phenotypic effect	Reference
SLG	S-locus/SLSG structural gene	self-incompatibility	18
SUP1	incompletely dominant suppressor of incompatibility in *B. oleracea*	reduction in SLSG self-compatibility	23
sup2	recessive suppressor of incompatibility in *B. campestris*	reduction in SLSG self-compatibility	this paper
mod1	recessive modifier of incompatibility in *B. campestris*	processing of SLSG? self-compatibility	24

One mechanism involves a drastic reduction of SLSG levels in the stigma. This mechanism is exemplified by the self-compatible *B. campestris* plants described in this paper and the previously reported *B. oleracea* var *capitata* self-compatible line. Experiments are under way to determine whether the reduction in SLSG occurs by decreased stability of SLSG, decreased stability of SLSG transcripts, or by down-regulation of the SLSG structural gene itself. Should the latter possibility be demonstrated, the implication would be that Sup1 and Sup2 encode trans-acting factors which specifically enhance the transcription of the SLSG structural gene in the stigma.

Another mechanism operates at the level of SLSG itself, and assumes that the expression of self-incompatibility is not only dependent on the presence of adequate levels of S-gene product in the stigma, but also on the correct modification/activation of the SLG primary translational product. The glycosylated and heterogeneous nature of SLSG (12) implies that SLSG undergoes post-translational modification. In this mechanism, the transcription of the S-locus is unaffected and SLSG is still produced. Self-compatibility results from mutations in genes which are responsible for the (*i*) post-translational modification of SLSG; (*ii*) activation of SLSG; or (*iii*) targeting of SLSG.

Thinking About Partial Self-Compatibility (Pseudo-Compatibility). Partial self-compatibility, where some pollen tubes escape the stigmatic barrier in self-pollinations or in crosses between plants with the same active S-allele, has been shown to be under complex genetic control (7). A basis for the occurrence of this partial compatibility in self-incompatible plants is suggested by the concepts outlined in this paper. Partial compatibility can be thought of as leakiness in the operation of the self-incompatibility system. This leakiness could result from somatic mutations or development changes that bring about a localized reduction in SLSG levels in individual papillar cells, or in sectors comprising a few cells. Pollen grains themselves may also be subject to similar perturbations in their ability to respond to the recognition molecules. In either case, the effect is the same: an escape from inhibition and growth of a pollen tube which may go on to form a zygote. The resulting progeny, like their parents, are self-incompatible since the gametic nuclei are unaffected.

Acknowledgments--The work described in this paper was supported by a grant from the U.S. Department of Energy.

LITERATURE CITED

1. BATEMAN AJ 1954 Self-incompatibility systems in angiosperms. II. *Iberis amara*. Heredity 8: 305-332
2. BATEMAN AJ 1955 Self-incompatibility systems in angiosperms. III. Cruciferae. Heredity 9: 52-68
3. HARUTA 1962 Studies on the genetics of self- and cross-incompatibility in cruciferous vegetables. (In Japanese, Eng. summary). Takii Plant Breed Exp Stn Res Bull 2: 1-169
4. HINATA K, T NISHIO 1978 Stigma proteins in self-incompatible *Brassica campestris* L. and self-incompatible relatives, with special reference to S-allele specificity. Jap J Genet 53: 27-33
5. HINATA K, K OKASAKI 1986 Role of the stigma in the expression of self-incompatibility in crucifers in view of genetic analysis. *In* DL Mulcahy, GB Mulcahy, E Ottaviano, eds, Biotechnology and Ecology of Pollen. Springer-Verlag, Berlin, pp 185-190
6. HINATA K, K OKASAKI, T NISHIO 1983 Gene analysis of self-incompatibility in *Brassica campestris* var yellow sarson (a case of recessive epistatic modifier). *In* Proc 6th Intern Rapeseed Conf, Paris, 1: 354-359
7. HODGKIN T 1980 The inheritance of partial self-compatibility in *Brassica oleracea* L. inbreds homozygous for different S-alleles. Theor Appl Genet 58: 101-106
8. KANDASAMY MK, D PAOLILLO, JB NASRALLAH, C FARADAY, ME NASRALLAH S-locus specific glycoproteins of *Brassica* accumulate in the cell wall of developing stigma papillae. Dev Biol (in press)
9. KHO YO, J BAER 1968 Observing pollen tubes by means of fluorescence. Euphytica 17: 298-302
10. LEWIS D Genetic versatility of incompatibility in plants. New Zealand J Bot 17: 637-644
11. NASRALLAH JB, TH KAO, ML GOLDBERG, ME NASRALLAH 1985 A cDNA clone encoding an S-locus specific glycoprotein from *Brassica oleracea*. Nature 318: 263-267
12. NASRALLAH JB, ME NASRALLAH 1984 Electrophoretic heterogeneity exhibited by the S-allele specific glycoproteins of *Brassica*. Experientia 40: 279-281
13. NASRALLAH JB, SM YU, ME NASRALLAH 1988 Self-incompatibility genes of *Brassica oleracea*: expression, isolation, and structure. Proc Natl Acad Sci USA 85: 5551-5555

14. NASRALLAH ME 1974 Genetic control of quantitative variation in self-incompatibility proteins detected by immunodiffusion. Genetics 76: 45-50
15. NASRALLAH ME, DH WALLACE 1967 Immunogenetics of self-incompatibility in *Brassica oleracea* L. Heredity 22: 519-527
16. NASRALLAH ME, DH WALLACE, RM SAVO 1972 Genotype, protein, phenotype relationships in self-incompatibility of *Brassica*. Genet Res 20: 151-160
17. OCKENDON DJ 1974 Distribution of self-incompatibility alleles and breeding structure of open-pollinated cultivars of Brussel sprouts. Heredity 33: 159-171
18. ROBERTS IN, AD STEAD, DJ OCKENDON, HG DICKINSON 1979 A glycoprotein associated with the acquisition of the self-incompatibility system by maturing stigmas of *Brassica oleracea*. Planta 146: 179-183
19. SAMPSON DR 1957 The genetics of self-incompatibility in the radish. J Heredity 48: 26-29
20. SAMPSON DR 1964 A one-locus self-incompatibility system in *Raphanus raphanistrum*. Can J Genet Cytol 6: 435-445
21. SEARS ER 1937 Cytological phenomena connected with self-sterility in the flowering plants. Genetics 22: 130-181
22. STOUT AB 1920 Further experimental studies on self-incompatibility in hermaphroditic plants. J Genet 9: 85-129
23. TATEBE T 1962 Studies on the genetics of self- and cross-incompatibility in Japanese radish, V. J Jap Soc Hort Sci 31: 185-192
24. THOMPSON KF 1957 Self-incompatibility in marrow-stem kale, *Brassica oleracea* var *acephala*. I. Demonstration of a sporophytic system. J Genet 55: 45-60
25. THOMPSON KF 1972 Competitive interaction between two S-alleles in a sporophytically-controlled incompatibility system. Heredity 28: 1-7
26. THOMPSON KF, JP TAYLOR 1971 Self-incompatibility in kale. Heredity 27: 459-471

Plant Reproduction: From Floral Induction to Pollination, *Elizabeth Lord and Georges Bernier*, Eds, 1989, The American Society of Plant Physiologists Symposium Series, Vol. I

MOLECULAR GENETICS OF SELF-INCOMPATIBILITY IN *BRASSICA*

JUNE B. NASRALLAH

Section of Plant Biology, Cornell University, Ithaca, NY 14853, USA

A number of events take place at the stigma surface of *Brassica* following pollination. In a successful pollination, pollen will first adhere to the papillar cell surface, hydrate, germinate, and then its pollen tube will digest its way into the walls of papillar cells (6) through the action of degrading enzymes. In an unsuccessful pollination, the first steps of pollen response do not take place; pollen germination and/or pollen tube growth are inhibited and the papillar cells exhibit a rejection response visualized as callose production (9). Thus, considerable interactions must occur between molecules of stigma and pollen origin, and these molecular interactions determine whether a papillar cell can be successfully invaded by a pollen tube.

While nothing is known of the molecular physiology of pollination events, substantial genetic information has been accumulated concerning the control of a specific interaction at the pollen-stigma interface, that of self-incompatibility. Self-incompatibility specificity in *Brassica* is genetically controlled by a single locus, the S-locus. This locus is highly polymorphic, with over 50 described alleles (20, 21), and codes in the stigma for polymorphic S-locus specific glycoproteins (SLSGs) (17, 19). In its simplest form, self-incompatibility, manifested in the abortion of self-pollen development and the failure of self-fertilization, operates whenever the same S-allele is present in stigma and pollen grain. However, complex allelic interactions at play in both stigma and pollen, and the action of unlinked modifier genes are known which affect the expression of the S-locus and the final incompatibility phenotype (M. Nasrallah, this volume, pp 146).

Our approach to elucidate the mechanism of self-incompatibility is based on the molecular genetic characterization of SLSGs and related molecules with the aim of determining their mode of action in self-recognition.

GENOMIC ORGANIZATION OF S SEQUENCES IN *BRASSICA*

DNA sequences encoding SLSG have been cloned and molecularly characterized (15, 16, 18). In contrast to the genetic prediction of a single gene involved in determining incompatibility specificity, blot analyses of *Brassica* genomic DNA probed with SLSG-cDNA reveal multiple bands of hybridization indicating the presence of several sequences related to the SLSG structural gene. The occurrence of these multiple sequences has been verified by the screening of *Brassica* genomic libraries with SLSG-cDNA. Starting with an S-allele homozygous line, we have isolated a number of clones all showing homology to SLSG-cDNA, but differing from this cDNA and from one another by their restriction maps and their nucleotide sequence (5, 18).

EXPRESSION OF S SEQUENCES

Using SLSG-cDNA as a probe, RNA blot analysis has shown that the expression of S sequences is temporally regulated. Transcripts with homology to the probe cannot be detected in leaf, seedling, or style tissue (16). They are, however, expressed in the stigma at levels which vary with the developmental stage of the flower buds. The development of *Brassica* flowering buds is sufficiently regular so as to allow their classification into groups of buds of equivalent developmental age. Reliable criteria for this classification are the size of the bud and its position along the inflorescence relative to the opened flowers. Anthesis, or flower opening, can be designated as "0", and each developmental stage can then be assigned a number referring to its distance in days from anthesis (Fig. 1). Stigmas are self-compatible in immature buds up to and including buds at 2 d prior to anthesis (stages to -2). They become self-incompatible at 1 d prior to anthesis (stage -1), and maximal levels of S-transcripts are also attained at this stage (15). *In situ* hybridization of pistil sections with ^{3}H-labeled cDNA probes, or with ^{35}S-labeled single-stranded RNA probes, has further shown that S transcripts are exclusively localized in the papillar cells of the stigma surface and cannot be detected in any other cell layer of the pistil (18).

Self-incompatibility sequences are expected to be expressed in one other tissue of the flower. The genetic control of self-incompatibility in *Brassica* implies that the self-recognition molecules must be expressed in the two interacting cells: in the papillar cells of the stigma as discussed in the previous section on the one hand, and in the pollen on the other hand. Differences in the timing of gene expression have been suggested to ultimately determine the genetic control of pollen-incompatibility phenotype, and thus provide the basis for the fundamental genetic difference between sporophytically-controlled and gametophytically-controlled plant incompatibility systems. In gametophytic systems, such as the one operating in *Nicotiana* where pollen-incompatibility phenotype is determined by the haploid genotype of the pollen grain, S-gene expression might be expected to occur in pollen after meiosis, or even after pollen germination. In sporophytically-determined systems, such as the *Brassica* system where pollen

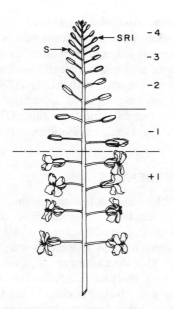

FIG. 1. Flower development along a *Brassica* inflorescence. Anthesis is indicated by an interrupted horizontal line, and the numbers refer to age of the buds and flowers in days from anthesis. The solid horizontal line separates the self-compatible zone in the younger buds from the self-incompatible zone in the more mature buds and flowers. The two arrows point to the bud stage in which S and SLR1 transcripts are first detected by *in situ* hybridization.

phenotype is determined by the diploid genotype of the pollen-producing plant, S-gene expression is expected (7), and is found (Nasrallah, in preparation), to occur in another tissue.

EXPRESSION OF INDIVIDUAL MEMBERS OF THE S-GENE FAMILY

It is evident, from the analysis of the expression of S sequences decribed in the previous paragraph, that however many members of the S-gene family are expressed, their expression is specific to the stigma and/or anther. We have so far shown that two gene members are in fact expressed. One of these genes is the SLSG structural gene, and is referred to as the SLG gene. The other gene has been designated SLR1 (12). The expression of both of these genes has been demonstrated by the isolation of corresponding clones from stigma cDNA libraries, and the use of the respective cDNA probes has revealed similarities and also profound differences in the properties of the two genes. A comparison of these properties is summarized below and is outlined in Table I.

The ease of obtaining cDNA sequences derived from SLG and SLR1 by differential screening of stigma cDNA libraries suggests that these two genes are the most abundantly-expressed S-genes in the stigma. Both genes show the same general pattern of expression. As determined by RNA blot analysis and by *in situ* hybridization, the transcripts of both genes attain maximal levels at 1 d prior to anthesis, and are exclusively localized in the stigma papillar cells. Differences in the details of expression can be noted, however. Based on their relative representation in cDNA libraries, SLR1 transcripts are approximately seven times more abundant than SLG transcripts. In addition, and perhaps as a consequence of their higher abundance, SLR1 transcripts are first detected by *in situ* hybridization a full day or several buds earlier than SLG transcripts (Fig. 1). Interestingly, stigmas also become competent in sustaining pollen tube growth at approximately the same early bud stage.

Table I. *Comparison of the SLG and SLR1 genes*

	SLG	SLR1[a]
Sequence data available for	several S-alleles in 2 species	several S-alleles in 2 species
Transcript localization	papillar cells	papillar cells
Protein localization	papillar cell wall	ND
Protein characteristics:		
1. Sequence divergence in different genotypes	highly variable	invariant
2. Predicted mol wt	46 kD	46 kD
3. Signal peptide	yes	yes
4. Cysteine residues	12	12
5. N-glycosylation sites	variable (8-10)	3
6. Molecular heterogeneity	yes	ND

[a]ND = not determined.

More substantial and significant differences between the two genes are revealed by sequence and genetic analyses. The conclusion that the SLG gene was involved in determining S-allele specificity was based largely on the demonstration of allele-associated sequence variability and on the cosegregation of the corresponding restriction fragment length polymorphisms with the S-alleles in genetic crosses. In comparisons between different S-allele homozygotes, it was shown that SLG sequences encoded polypeptides that were, at most, 80% homologous in their most conserved regions, and only 40% homologous in a variable region which we suggested to be a determinant of allelic specificity (15). Extensive polymorphism and cosegregation with S-alleles has been documented for the S-locus specific glycoproteins themselves (8, 17), and for the S-related DNA sequences revealed by hybridization with SLSG-encoding cDNA probes (16).

In contrast, the analysis of SLR1 sequences isolated from three different S-allele homozygotes has demonstrated that this gene is highly conserved and encodes identical proteins in strains that differ in the S-alleles, and cannot therefore be a determinant of allelic specificity. This high degree of conservation is reflected in the limited restriction fragment length polymorphism exhibited by SLR1 sequences in a survey of a number of different

S genotypes. A clear case of SLR1-associated polymorphism was, however, uncovered in a comparison of two lines of *B. oleracea*, a kale inbred homozygous for the S_6 allele and a cabbage inbred homozygous for the S_{14} allele. This polymorphism was used to demonstrate that the SLR1 gene lies outside the S-locus (12). Figure 2 shows an example of this analysis as performed on the F_2 progeny of a cross between the S_6 and the S_{14} inbreds. Genomic DNA was prepared from the S_6 and the S_{14} parents, and from 13 F_2 plants, restricted with EcoRI, fractionated on agarose gels and transferred to Gene Screen Plus. The filter was hybridized first with an SLG probe (top panel) and then with an SLR1 probe (lower panel), as described earlier (5). A comparison of the restriction patterns obtained with the two probes for each plant clearly shows that the two genes are unlinked. For example, plant #1, which is heterozygous for the SLG gene, exhibits only the SLR1 restriction fragment characteristic of the kale parent and is, therefore, homozygous at the SLR locus; and plant #6, which is homozygous at the SLG locus, exhibits both the SLR1 restriction fragments characteristic of the cabbage parent and

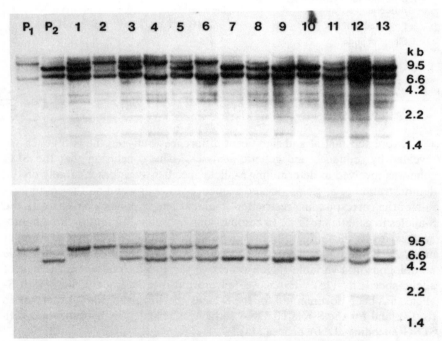

FIG. 2. Analysis of a population of *B. oleracea* plants segregating for the S_6 and S_{14} alleles. DNA was isolated from the S_6 homozygous parent (P_1), the S_{14} homozygous parent (P_2), and 13 F_2 progeny plants. A blot of EcoRI-digested DNA was hybridized with the SLG gene probe (top panel) and with the SLR1 gene probe (lower panel). Molecular weight markers are indicated.

of the kale parent and is, therefore, heterozygous at the SLR1 locus. More recently, we have also analyzed an SLR1 polymorphism in *B. campestris*. The results, shown in Figure 3, again demonstrate the independent segregation of the two genes and that, in this species also, the SLR1 gene lies outside the S-locus.

DISCUSSION

We initiated our investigations into the structure of the SLG gene with the aim of understanding the molecular basis of allelic specificity in pollen recognition. This study has led us to the identification in the *Brassica* genome, of a family of related genes which appear to be involved in processes that operate at the pollen-stigma interface.

The existence of the SLR1 gene had not been revealed by classical genetic studies. The demonstration that this additional member of the S-gene family is also expressed, suggests the existence of a family of related proteins in the *Brassica* flower. The primary translation products of SLG and SLR1 are 70% homologous and share common structural features which may be the hallmark of the self-incompatibility gene family products. These features include the presence of a signal peptide (suggesting that these molecules may have similar subcellular localization), glycosylation (SLSGs are glycosylated and the putative SLR1 product is predicted to be glycosylated), and particularly the same precise arrangement of 12 cysteine residues at the carboxy terminus of the molecule.

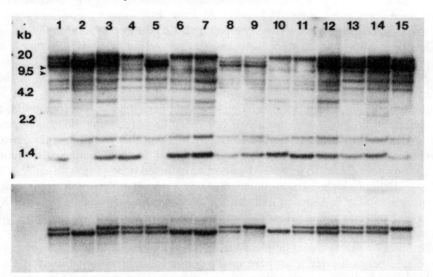

FIG. 3. Analysis of a population of *B. campestris* plants segregating for two S-alleles. A blot of EcoRI-digested DNA was hybridized with the SLG gene probe (top panel) and with the SLR1 gene probe (lower panel). The two arrowheads in the top panel indicate the positions of the two SLR1 bands. Molecular weight markers are indicated.

An intriguing question concerns the function of these proteins and their involvement in pollen recognition. The obvious differences in selective pressure operating on SLG and SLR1, to generate diversity in the first case and to maintain identity in the second case, must be related to the very different biological functions of the two gene products. While molecular data, together with extensive genetic evidence, has assigned the SLG gene and its SLSG product a central role in the determination between "self" and "nonself" pollen, the predicted SLR1 glycoprotein has not been identified and nothing is known of its possible role. The extreme conservation of SLR1 sequences even in different *Brassica* species, and their expression to high levels even in self-compatible strains in which the SLG gene is either not functional or not expressed (12), suggests for the SLR1 protein product a fundamental role in pollination events in *Brassica*. A model which suggests the evolution of self-incompatibility by successive mutations to active incompatibility alleles from an ancestral compatibility allele has been recently presented (4). It is thus tempting to speculate that SLR1 may represent the ancestral gene from which the other members of the S-gene family arose by duplication. It may be assumed that at least one of these duplicated copies, freed from the selective forces operating on SLR1, diverged and became the first functional self-incompatibility gene.

The complexity of recognition phenomena is well documented. One of the best studied examples is cellular immunity in mammals. Its control by the major histocompatibility complex (MHC) and plant self-incompatibility systems have been likened because of their common ability to discriminate between "self" and "nonself" (3, 13). At least three major families of molecules are encoded by this large complex genetic region, and immune function has been shown to be dependent on the interaction of MHC glycoprotein products with a number of related and unreleated molecules (10, 11). Two examples of such interactions are particularly relevant in the context of the expression of the unlinked nonpolymorphic SLR1 gene in *Brassica*. One is the association of the polymorphic class I antigens with β_2-microglobulin, a conserved molecule with sequence homology to histocompatibility antigens, but encoded by a gene unlinked to the MHC. Another example is the intracellular association of the polymorphic class II antigens with an invariant glycoprotein chain (li) encoded by another unlinked gene (14).

Recent data have underlined several parallels between self-incompatibility in *Brassica* and cellular immunity in mammals. Noteworthy in both systems are the involvement of highly polymorphic glycoproteins, the high degree of allele-encoded DNA sequence variability, genetic control by a complex locus consisting of a cluster of related genes, and the occurrence of related genes located elsewhere in the genome. Self-incompatibility is also likely to involve a complex series of molecular interactions, and it is tempting to draw a parallel with the MHC and suggest that, like the invariant li chain or β_2-microglobulin, the nonpolymorphic SLR1 gene product may be associated with SLSG in self-incompatible strains. In any case, the final incompatibility phenotype can be expected to be the result of the action of several different genes. Some of these genes may be expressed members of the S-multigene family, while others may be unrelated genes which affect the functioning of the self-incompatibility system.

It is likely that gametophytic incompatibility systems will also prove to involve a complex series of molecular interactions. Polymorphic S-allele associated glycoproteins and their corresponding cDNAs have been identified in *Nicotiana alata* (1). No homology can, however, be found between the *Brassica* and *Nicotiana* molecules either at the DNA sequence level, or at the level of predicted overall protein structure. Furthermore, and in marked contrast to the situation in *Brassica*, only one gene with homology to the S-associated cDNAs apparently exists in the *Nicotiana* genome (2). It is, therefore, not yet evident how general the features of the Brassica self-incompatibility system will turn out to be.

Acknowledgments--I wish to acknowledge Dr. S. M. Yu for performing the *in situ* hybridization on developing stigmas. The work described in this paper was supported by a grant from the U.S. Department of Agriculture.

LITERATURE CITED

1. ANDERSON MA, EC CORNISH, SL MAU, EG WILLIAMS, R HOGGART, A ATKINSON, I BONIG, B GREGO, R SIMPSON, PJ ROCHE, JD HALEY, JD PENSCHOW, HD NIALL, GW TREGEAR, JP COCHLAN, RJ CRAWFORD, AE CLARKE 1986 Cloning of a cDNA for a stylar glycoprotein associated with the expression of self-incompatibility in *Nicotiana alata*. Nature 321: 38-44
2. BERNATZKY R, MA ANDERSON, AE CLARKE 1988 Molecular genetics of self-incompatibility in flowering plants. Dev Genet 9: 1-12
3. BURNET FM 1971 Evolution of the inmune process in vertebrates. Nature 218: 426-430
4. CHARLESWORTH D 1988 Evolution of homomorphic sporophytic self-incompatibility. Heredity 60: 445-453
5. DWYER KG, A CHAO, B CHENG, CH CHEN, JB NASRALLAH 1989 The *Brassica* self-incompatibility multigene family. Genome (in press)
6. ELLEMAN CJ, CE WILLSON, RH SARKER, HG DICKINSON 1988 Interaction between the pollen tube and stigmatic cell wall following pollination in *Brassica oleracea*. New Phytol 109: 111-117
7. HESLOP-HARRISON J 1968 Pollen wall development. Science 161: 230-237
8. HINATA K, T NISHIO 1978 Stigma proteins in self-incompatible *Brassica campestris* L. and self-incompatible relatives, with special reference to S-allele specificity. Jap J Genet 53: 27-33
9. KANNO T, K HINATA 1969 An electron microscopic study of the barrier against pollen-tube growth in self-incompatible Cruciferae. Plant Cell Physiol 10: 213-216

10. KAUFMAN JF, C AUFFRAY, AJ KORMAN, DA SCHACKELFORD, J STROMINGER 1984 The class II molecules of the human and murine major histocompatibility complex. Cell 36: 1-13
11. KLEIN J 1986 Natural history of the major histocompatibility complex. John Wiley and Sons, New York
12. LALONDE B, ME NASRALLAH, KG DWYER, CH CHEN, B BARLOW, JB NASRALLAH 1989 A highly conserved *Brassica* gene with homology to the S-locus specific glycoprotein structural gene. The Plant Cell 1: 249-258
13. LARSEN K 1986 Cell-cell recognition and compatibility between heterogenic and homogenic incompatibility. Hereditas 105: 115-133
14. LONG EO, M STRUBIN, CT WAKE, N GROSS, S CARREL, P GOODFELLOW, RS ACCOLLA, B MACH 1983 Isolation of cDNA clones for the p33 invariant chain associated with HLA-DR antigens. Proc Natl Acad Sci USA 80: 5714-5718
15. NASRALLAH JB, TH KAO, CH CHEN, ML GOLDBERG, ME NASRALLAH 1987 Amino-acid sequence of glycoproteins encoded by three alleles of the S-locus of *Brassica oleracea*. Nature 326: 617-619
16. NASRALLAH JB, TH KAO, ML GOLDBERG, ME NASRALLAH 1985 A cDNA clone encoding an S-locus specific glycoprotein from *Brassica oleracea*. Nature 318: 263-267
17. NASRALLAH JB, ME NASRALLAH 1984 Electrophoretic heterogeneity exhibited by the S-allele specific glycoproteins of *Brassica*. Experientia 40: 279-281
18. NASRALLAH JB, SM YU, ME NASRALLAH 1988 Self-incompatibility genes of *Brassica oleracea*: expression, isolation and structure. Proc Natl Acad Sci USA 85: 5551-5555
19. NISHIO T, K HINATA 1982 Comparative studies on S-glycoproteins purified from different S-genotypes in self-incompatible *Brassica* species. I. Purification and chemical properties. Genetics 100: 641-647
20. OCKENDON DJ 1982 An S-allele survey of cabbage (*Brassica oleracea* var. *capitata*). Euphytica 31: 325-331
21. THOMPSON JF 1957 Self-incompatibility in marrow-stem kale, *Brassica oleracea* var. *acephala*. 1. Demonstration of a sporophytic system. J Genet 55: 45-60

MOLECULAR GENETICS OF SELF-INCOMPATIBILITY IN FLOWERING PLANTS

Marilyn A. Anderson, Paul R. Ebert, Mitchell Altschuler,
and Adrienne E. Clarke

Plant Cell Biology Research Centre, School of Botany, University of Melbourne, Parkville, Victoria 3052, Australia

Flowering plants have evolved a number of mechanisms to ensure outcrossing and prevent inbreeding. Among these, the most widespread is the genetically controlled trait of homomorphic self-incompatibility which is defined as "the inability of female hermaphrodite seed plants to produce zygotes after self-pollination" (22). The mechanism for self-incompatibility is controlled by the self-incompatibility (*S*) gene which prevents fertilization between gametes produced by the same plant and different plants bearing the same alleles of the *S*-gene. This process has been the subject of close study by biologists for over a century. For example, Darwin described self-incompatibility and its role in ensuring outcrossing in the evolution of flowering plants in 1877 (9). The early biologists described the biology and genetics of the phenomena in such detail that by the mid 1980s, there were several experimental systems available which could be dissected at the molecular level using recombinant DNA techniques. In this review, we will consider the current state of knowledge of the molecular genetics of self-incompatibility in the two major systems of homomorphic self-incompatibility. There has been intense interest in this subject during the last few years not only because of its intrinsic importance in the field of pollination biology, but also because it is a relatively simple example of recognition and response in higher plants. It seems likely that an understanding of the molecular mechanisms of self recognition will lead to principles which may be applicable to other systems involving recognition between genetically dissimilar partners, such as pathogenic or symbiotic interactions between fungi and higher plants. Because of this current interest, we have recently reviewed the subject on several occasions (3, 6-8, 11) and will draw on the material presented in these publications.

BIOLOGY AND GENETICS OF SELF-INCOMPATIBILITY

The female organ of flowering plants is the pistil, which consists of the stigma, style, and ovary. The male gametes are carried in pollen which is transported to the stigma surface by vectors such as insects or wind. In a compatible mating, pollen hydrates and germinates, producing a tube, which grows through the style to the embryo sac where the two sperm cells are released; one is involved in fertilization and the other fuses with the primary endosperm nucleus. The pistil can discriminate between different types of pollen; usually wide intergeneric and interspecific crosses are prevented, while intraspecific crosses are successful except when self-incompatibility genes prevent inbreeding.

There are two major self-incompatibility systems, heteromorphic and homomorphic. Details of the biology and genetics of these systems are given in reviews by Lewis (19), Pandey (29), Linskens (20), and de Nettancourt (10). Heteromorphic plants of the same species characteristically produce morphologically distinct flowers. The distinctive features of the flower involved in the control of self-incompatibility are the length of the style and its relation to the level of the anthers. There have not been any molecular studies of heteromorphic self-incompatibility, and it is not discussed further in this review. Two useful reviews of the subject have been presented by Ganders (13) and Gibbs (14).

There are two types of homomorphic self-incompatibility--gametophytic, in which the haploid pollen expresses its own S-genotype, and sporophytic, in which pollen behavior is determined by the genotype of the pollen producing plant.

Gametophytic Self-Incompatibility. This system is the most common and occurs, for example, in taxa of the Solanaceae, Liliaceae, Poaceae, Commelinaceae, Onagraceae, Papaveraceae, Rosaceae, and Rubiaceae. In most systems, the specificity of the self-incompatibility response is governed by a single S gene with multiple alleles. There may be 40 or more different alleles in a particular system (18). In gametophytic systems, a pollen tube carrying a single S-allele is inhibited if the same allele is present in the style. Fertilization is successful only if the S-allele carried by the pollen is different from either of the two carried by the diploid tissues of the pistil. Although many gametophytic self-incompatibility systems are controlled by the action of a single multiallelic gene, there are more complex systems involving two or more loci which behave independently (17, 21, 28).

In many species, compatible and incompatible pollen appear to hydrate and germinate normally. As the tubes grow through the style, deposits of callose, detected by its staining with the aniline blue fluorochrome (12, 31), are deposited at regular intervals giving the tubes a ladder-like appearance. The callose is believed to form a barrier between the cytoplasm and the sperm cells, which are contained in the tube tip, and the spent pollen grains. At some point during the growth of the incompatible pollen tube, its appearance becomes different from that of compatible tubes and growth ceases. Characteristically, the tube wall becomes thickened and the tube tips often swell and may burst. There is often a deposit of callose close to the tip. This pattern is typical of self-incompatibility in the Solanaceae and the Rosaceae, while in grasses, growth of incompatible pollen tubes is arrested either at the stigma surface or very close to the surface (16).

Sporophytic Self-Incompatibility. This system is not as widespread as the gametophytic system and has only been well described for two families, the Cruciferae (Brassicaceae) and the Compositae (Asteraceae). The genetic model for this system is based on a single, multi-allelic S-locus. The incompatibility reaction is determined by the alleles carried by the pollen-producing parent plant, rather than the S-allele of the pollen itself. The S-alleles may express dominance relationships or independence in the pollen or pistil (33-35) which is in contrast to the gametophytic system in which the S-alleles act independently in both the pollen and pistil.

In sporophytic systems, the growth of incompatible pollen tubes is arrested close to the stigma surface. In some cases, the pollen fails to germinate. This is in contrast to the most common case for gametophytic systems in which tube arrest is within the tissues of the style. Callose deposition is also associated with tube arrest in the sporophytic systems and the deposits appear in the tip of the emerging pollen tube and at the site of contact on the papilla cell.

Developmental Regulation of the S-Gene. Self-incompatibility is not expressed in immature pistils of plants having either gametophytic or sporophytic self-incompatibility systems. In both systems, self-incompatibility is usually first expressed in pistils about 24 h before the flower opens. Thus, immature pistils are self-compatible, and homozygous plants can be generated by self-pollination of immature buds. The ability to generate homozygous material in this way has been an important tool for studies of the molecular genetics of self-incompatibility.

IDENTIFICATION OF S-GENE PRODUCTS IN FEMALE SEXUAL TISSUES

During the 1970s and early 1980s proteins were identified in the mature pistils of a number of self-incompatible species which corresponded to particular S-alleles (for review, see 7, 8). These proteins were subsequently shown to be glycosylated and were referred to as "S-allele associated glycoproteins" (2). Important studies by Nasrallah *et al.* (23) and Nishio and Hinata (27) in *Brassica* species showed that not only did a particular glycoprotein correspond to a particular S-allele, but that it segregated with S-allele function through two generations. This indicated that the glycoprotein was either the product of the S-allele or of a gene closely linked to it. Since then, several studies on other self-incompatible species have demonstrated similar segregation of S-allele function with particular stigma or style glycoproteins. The best studied of these glycoproteins are those from stigmas of *B. campestris* and *B. oleracea* (sporophytic) and *Nicotiana alata* (gametophytic).

MOLECULAR GENETICS OF THE GAMETOPHYTIC S-LOCUS OF *NICOTIANA ALATA*

The initial information on the molecular genetics of self-incompatibility in *N. alata* came through the isolation of cDNA clones encoding the style glycoproteins which segregate with the $S2$, $S3$, and $S6$ alleles. The first cDNA clone ($S2$ allele) was obtained by differential screening and the identity confirmed using amino acid sequence information obtained from the isolated $S2$ style glycoprotein (2). The other cDNAs encoding the $S3$ and $S6$ allele were obtained by differential screening (1) and their identity confirmed by homology with the first clone. These cDNAs were used as probes on Southern blot analyses of genomic DNA from plants carrying the $S1$, $S2$, $S3$, $S6$, and $S7$ alleles (1, 4). Several restriction enzymes were used and, in most cases, single fragments hybridizing with the probes were obtained. The fragments produced were only large enough to contain a single copy of the coding sequence, indicating that the system is a single locus, single copy gene system. As well as the major hybridizing fragments, weakly hybridizing secondary bands were also obtained; the nature and significance of these bands is not understood.

There were two other interesting features of the Southern analysis, one was the abundance of restriction fragment length polymorphisms between alleles and the second was the weak cross hybridization between alleles. These features indicate variability both within the coding regions and in the flanking regions. The sequence data for the cDNAs encoding the $S2$, $S3$, and $S6$ alleles confirms this, as there is only 63 to 70% amino acid identity between the various pairs of alleles, and only 51% of the amino acids are conserved between the three alleles (Fig. 1). Although there are amino acid substitutions, deletions, and insertions throughout the sequences, there are 10 cysteine residues which are conserved between alleles. (The $S2$-sequence lacks one.) There are four potential glycosylation sites which are conserved between the three alleles. (The $S3$-sequence has one additional site.) The conservation of the cysteine residues and glycosylation sites suggests that these are essential structural features of the S-allele products, although it is possible that the glycosyl residues may also contribute to allelic specificity.

A recent interesting finding is that there is a short sequence of 56 bp, located 481 bp upstream of the transcription initiation site of the S-gene, which is homologous with a mitochondrial sequence. The nuclear sequence is marked at the 5' end by two 8 bp direct repeats and at the 3' end by an inverted 7 bp repeat. The origin and possible functional significance of this sequence is not yet understood (5).

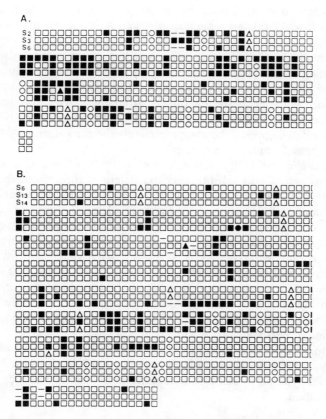

FIG. 1. Comparison of fully-sequenced S-alleles. (A) Comparison of the amino acid sequence of three N. alata alleles, S2, S3, and S6, derived from the nucleotide sequence of their respective cDNAs and from amino terminal peptide sequence (2). (B) Comparison of the derived amino acid sequences of three cloned B. oleracea cDNAs, S6, S13, and S14 (25). Unfilled symbols refer to amino acids which are conserved between two or three alleles, while filled symbols denote amino acids which are unique to one allele at that particular location. (O) denotes a cysteine residue, (Δ) denotes the asparagine residue of a potential N-linked glycosyl chain attachment site, while (-) denotes a gap introduced into the sequence to provide optimal alignment between alleles. All other amino acids are represented by a (□). Reproduced from (11) with permission of Cell Press.

MOLECULAR GENETICS OF THE SPOROPHYTIC S-LOCUS OF
BRASSICA OLERACEA

Three cDNA clones corresponding to different S-alleles of B. oleracea have been isolated and sequenced (24-26). The amino acid sequences of pairs of these cDNAs showed 79 to 85% identity. However, in contrast to the gametophytic system of *N. alata*, the *B. oleracea* S-gene system consists of a family of at least 11 related sequences of which only one per haploid genome is active (26).

The majority of the amino acid sequence of the stigma glycoprotein corresponding to the $S8$ allele of *B. campestris* was determined directly from the isolated glycoprotein by Takayama and coworkers (32). Comparison of this sequence with the derived sequence from the cDNA encoding the $S6$ allele of *B. oleracea* showed 82% identity. Thus, the alleles from two distinct but closely related species show approximately the same extent of homology as do two alleles from a single species.

The allelic variability between the S-alleles of *Brassica* species is expressed as amino acid substitutions and deletions throughout the primary sequence (Fig. 1). There are 12 cysteine residues which are conserved in the carboxyl domain of the products of the four alleles of the *Brassica* species. As for the *N. alata* sequence, there are conserved glycosylation site concensus sequences. Direct amino acid sequencing (32) indicates that of nine potential glycosylation sites in *B. campestris*, seven are glycosylated and five of these sites are conserved among all three of the *B. oleracea* alleles. As for the *N. alata* system, the role that these glycosyl residues might play in structure or allelic specificity is not clear. A recent report on the effect of the glycosylation inhibitor tunicamycin on excised styles suggests a possible role in determining specificity (30).

Comparison of the amino acid sequences of the S-alleles of *N. alata* and *B. oleracea* shows that there is no homology; this leads to the suggestion that the two systems of self-incompatibility may have arisen independently during evolution.

THE NEXT QUESTIONS

The cloning of cDNAs corresponding to the style glycoproteins which segregate with particular S-alleles, has lead to the isolation of corresponding genomic clones and to transformation studies. Details of these studies are not yet published, but they will lead to insights into generation of allelic diversity and control of S-gene action. The idea that gene conversion mechanisms, similar to those implicated in the generation of allelic diversity in animal gene systems, might be involved in the generation of diversity between S-alleles in the two major types of self-incompatibility, has been explored recently (11).

A major new challenge is to define the corresponding allelic product in pollen. Our experience is that there is no protein which corresponds with a particular S-allele which can be detected in 2-dimensional protein gels of pollen extracts. However, as the gene is expressed in both pollen and style, it is assumed that there is a protein product which may be expressed at very low levels or in a very restricted part of the pollen tube. When the nature of the pollen product of the S-gene has been identified the next questions of how the products of a particular allele in the pollen and style interact to cause arrest of pollen tube growth can be addressed. The finding that isolated *S2* glycoprotein from *N. alata* styles inhibits pollen tube growth *in vitro* with some specificity (15), suggests that there may well be a direct interaction of style *S*-allele product with the pollen tube *in vivo*. This leads to the question of how the signal from the style is received at the pollen tube--Does the wall play an active role? Does the signal move through the wall of the pollen tube to the plasma membrane? What is the nature of the receptor? How is the signal transduced and what is the nature of the messenger which ultimately leads to a disruption of wall biosynthesis? How is the deposition of callose controlled and what is its role? We have available many studies of recognition and response in animal cell systems on which we can model our approaches to these questions, but we do not have corresponding studies and experience in plant cell recognition systems to guide us. It may well be that the cell wall, which is one of the key differences between animal and plant cells, plays an essential role in the sequence of events which comprise the self recognition during expression of self-incompatibility.

LITERATURE CITED

1. ANDERSON MA, A ATKINSON, R BERNATZKY, T ORPIN, H DEDMAN, G TREGEAR, R FERNLEY, AE CLARKE 1989 Sequence variability of three alleles of the self-incompatibility gene of *Nicotiana alata*. (Submitted)
2. ANDERSON MA, EC CORNISH, S-L MAU, EG WILLIAMS, R HOGGART, A ATKINSON, I BONIG, B GREGO, R SIMPSON, PJ ROCHE, JD HALEY, HD NIALL, GW TREGEAR, JP COGHLAN, RJ CRAWFORD, AE CLARKE 1986 Cloning of cDNA for a stylar glycoprotein associated with expression of self-incompatibility in *Nicotiana alata*. Nature 321: 38-44
3. BACIC A, AE CLARKE 1988 Molecular approaches to understanding cellular recognition in plants. *In* JL Key, L McIntosh, eds, The Molecular Basis of Plant Development. UCLA Symp Mol Cell Biol. Alan R Liss, New York, pp 53-66
4. BERNATZKY R, MA ANDERSON, AE CLARKE 1988 Molecular genetics of self-incompatibility in flowering plants. Dev Genet 9: 1-12
5. BERNATZKY R, S-L MAU, AE CLARKE 1989 A nuclear sequence associated with self-incompatibility in *Nicotiana alata* has homology with mitochondrial DNA. Theor Appl Genet (in press)
6. CLARKE AE, MA ANDERSON, A BACIC, EC CORNISH, PJ HARRIS, S-L MAU, JR WOODWARD 1987 Molecular aspects of self-incompatibility in flowering plants. *In* JL Key, L McIntosh, eds, Plant Gene Systems and Their Biology. Alan R Liss, Inc, New York, pp 53-64
7. CORNISH EC, MA ANDERSON, AE CLARKE 1988 Molecular aspects of fertilization in flowering plants. Annu Rev Cell Biol 4: 209-228
8. CORNISH EC, JM PETTITT, AE CLARKE 1988 Self-incompatibility genes in flowering plants. *In* DPS Verma, R Goldberg, eds, Temporal and Spatial Regulation of Plant Genes, Vol 5. Springer-Verlag, New York, pp 117-130

9. DARWIN C 1877 The different forms of flowers on plants of the same species. John Murray, London
10. DE NETTANCOURT D 1977 Incompatibility in Angiosperms. Springer-Verlag, Berlin
11. EBERT PR, MA ANDERSON, R BERNATZKY, M ALTSCHULER, AE CLARKE 1989 Genetic polymorphism of self-incompatibility in flowering plants. Cell 56: (in press)
12. EVANS NA, PA HOYNE, BA STONE 1984 Characteristics and specificity of the interaction of a fluorochrome from aniline blue (Sirofluor) with polysaccharides. Carbohydr Polym 4: 215-230
13. GANDERS FR 1979 The biology of heterostyly. NZ J Bot 17: 607-635
14. GIBBS P 1986 Do homomorphic and heteromorphic self-incompatibility systems have the same sporophytic mechanisms? Plant Syst Evol 154: 285-323
15. HARRIS PJ, JA WEINHANDL, AE CLARKE 1989 Effect on *in vitro* pollen growth of an isolated style glycoprotein associated with self-incompatiblility in *Nicotiana alata*. Plant Physiol 89: (in press)
16. HESLOP-HARRISON J 1982 Pollen-stigma interaction in the grasses: a brief review. NZ J Bot. 17: 537-546
17. LARSEN K 1977 Self-incompatibility in *Beta vulgaris* L. I. Four gametophytic, complementary S-loci in sugar beet. Hereditas 85: 227-248
18. LAWRENCE MJ 1975 The genetics of self-incompatibility in *Papaver rhoeas*. Proc R Soc Lond B 188: 275-285
19. LEWIS D 1949 Incompatibility in flowering plants. Biol Rev 24: 472-496
20. LINSKENS HF 1960 Zur Frage der Entstehung der Abwehr-Korpen bei der Inkompatibilittsreaktion von Petunia. III Mittelung: Serologische Teste mit Leitgewebs und Pollen-Extrakten. Z Bot 48: 126-135
21. LUNDQUIST A 1964 The nature of the two-loci incompatibility system in grasses. IV. Interaction between the loci in relation to pseudo-compatibility in *Festuca pratensis* Huds. Hereditas 52: 221-234
22. LUNDQUIST A, U OSTERBYE, K LARSEN, I LINDE-LAURSEN 1973 Complex self-incompatibility systems in *Ranunculus acris* L. and *Beta vulgaris* L. Hereditas 74: 161-168
23. NASRALLAH ME, JT BARBER, DH WALLACE 1970 Self-incompatibility proteins in plants: detection, genetics and possible mode of action. Heredity 25: 23-27
24. NASRALLAH JB, T-H KAO, C-H CHEN, ML GOLDBERG, ME NASRALLAH 1987 Amino-acid sequence of glycoproteins encoded by three alleles of the S locus of *Brassica oleraceae*. Nature 326: 617-619
25. NASRALLAH JB, T-H KAO, ML GOLDBERG, ME NASRALLAH 1985 A cDNA clone encoding an S locus-specific glycoprotein from *Brassica oleraceae*. Nature 318: 263-267
26. NASRALLAH JB, S-M YU, ME NASRALLAH 1988 Self incompatibility genes of *Brassica oleraceae* expression, isolation, and structure. Proc Natl Acad Sci USA 85: 5551-5555
27. NISHIO T, K HINATA 1978 Stigma proteins in self-incompatible *Brassica campestris* L. and self-compatible relatives, with special reference to S allele-specificity. Jpn J Genet 53: 27-33
28. OSTERBYE U 1975 Self-incompatibility in *Ranunculus acris* L. Genetics, interpretation and evolutionary aspects. Hereditas 80: 91-112

29. PANDEY KK 1958 Time of S-allele action. Nature 181: 1220-1221
30. SARKER RH, CJ ELLEMAN, HG DICKINSON 1988 Control of pollen hydration in *Brassica* requires continued protein synthesis, and glycosylation is necessary for intraspecific incompatibility. Proc Natl Acad Sci USA 85: 4340-4344
31. STONE BA, NA EVANS, I BONIG, AE CLARKE 1984 The application of Sirofluor, a chemically defined fluorochrome from aniline blue, for the histochemical detection of callose. Protoplasma 122: 191-195
32. TAKAYAMA S, A ISOGAI, C TSUKAMOTO, Y UEDA, K HINATA, K OKAZAKI, A SUZUKI 1987 Sequences of the S-glycoproteins, products of the *Brassica campestris* self-incompatibility locus. Nature 326: 102-105
33. THOMPSON KF 1957 Self-incompatibility in marrow-stem kale, *Brassica oleracea* var. *acephala*. I. Demonstration of a sporophytic system. J Genet 55: 45-60
34. THOMPSON KF 1972 Competitive interaction between two S alleles in a sporophytically-controlled incompatibility system. Heredity 28: 1-7
35. THOMPSON KF, SP TAYLOR 1966 Non-linear dominance relationships between S-alleles. Heredity 21: 345-362

INTERACTION BETWEEN MUTATIONS THAT INFLUENCE INFLORESCENCE DEVELOPMENT IN MAIZE

Bruce Veit and Sarah Hake

Plant Gene Expression Center, 800 Buchanan Street, Albany, CA 94710, USA

We have begun to analyze the extent to which different mutations affecting inflorescence development interact with each other. Of several hundred mutations that have been isolated in corn, some twenty have relatively specific effects on various aspects of inflorescence development. By considering the phenotypes of each of these mutations, and the extent to which they are modified by the presence of additional inflorescence-specific mutations, we hope to gain insight into the functional relationships between genes involved in inflorescence development.

Initial experiments indicate that the tassel morphology conditioned by *ts2* (*ts* = tasselseed) is modified by *Mp1* (miniplant) or *ts4*. *ts2* typically produces a tassel in which the normally staminate florets are instead completely pistillate. *Mp1* conditions both dwarf and anther-ear phenotypes, the latter involving the developlent of anthers in the kernel-bearing florets of the ear. In the *Mp1 ts2* double mutant, both the dwarf and anther-ear phenotypes persist. Interestingly, the pistillate tassel florets conditioned by *ts2* become perfect when *Mp1* is also present. Thus, in its response to *Mp1*, the pistillate tassel inflorescence produced by *ts2* behaves much like its normal axillary counterpart, the ear. *ts4* conditions the development of a highly branched tassel with the sporadic development of pistillate florets. In *ts2 ts4* double mutants, a highly branched, completely pistillate tassel develops.

With both the *ts2 Mp1* and *ts2 ts4* double mutants, the mutant phenotypes can be interpreted as simple superimpositions of the transformations associated with each of the single mutants (*i.e.* the elaboration of a given mutant phenotype is not precluded by the presence of the second mutation). It would seem that these mutations affect developmental processes that are relatively distinct from each other.

In contrast, *Ts6* and *Tu* (tunicate) seem to interact synergistically to effect transformations not seen in either single mutant. By itself, *Ts6* produces a highly branched and completely pistillate tassel, but has relatively little effect on ear development. *Tu* conditions the exaggerated development of glumes, leaf-like structures that form part of maize florets in both the ear and tassel. *Ts6 TU* double mutant plants bear extremely branched ears and tassels which, interestingly, completely lack glumes. The inhibition of *Tu* conditioned glume development by *Ts6* suggests these two genes affect interdependent developmental processes.

THE *ms1* LOCUS IN SOYBEAN: ORIGIN OF MALE AND FEMALE GAMETES

Fan Zhang, Reid G. Palmer, and Halina Skorupska

Departments of Agronomy and genetics, USDA-ARS, Iowa State University, Ames, IA 50011, USA

The nuclear male-sterile (*ms1*) gene in soybean (*Glycine max* [L.] Merr.) is inherited monogenetically. Seven independent spontaneous mutations at the *ms1* locus have been identified. Homozygous recessive *ms1 ms1* plants are characterized by formation of coenocytic microspores due to failure of cytokinesis after telophase II of meiosis. The *ms1* locus is associated with reduced female fertility and with increased frequencies of polyembroyony, polyploidy, and haploidy in the progeny of sterile plants. Cytological investigations of *ms1 ms1* ovules have uncovered aberrant megagametophytes, though the number of mitotic divisions of megaspores was not affected by the *ms1* gene. Development of all four nuclei may lead to mature megagametophytes with up to four times the normal number of nuclei. Gamete function of *ms1ms1* genotype was analyzed using the *y11* codominant marker, a chlorophyll deficient mutation. In F_2 populations, segregation at the *Y11 y11* locus is 1 green:2 yellow-green:1 yellow lethal phenotypes.

A test for apomixis was conducted to determine the reason for abnormalities in the progeny of *ms1 ms1* plants. During 1987 and 1988, 5929 emasculations were made on flowers of *ms1 ms1 Y11 y11* plants grown in the greenhouse in isolation. No seeds were obtained. In this experiment, no evidence was found to support the occurrence of apomixis in *ms1 ms1* soybean. Variation in chromosome number of progeny of *ms1 ms1* plants cannot be explained by apomixis.

In the second experiment, we analyzed the progeny of nonemasculated *ms1 ms1 Y11 y11* plants grown in isolation. One hundred twenty-seven seeds were harvested and tested for chromosome number and plant color. These plants will also be checked for fertility at anthesis. Seed on *ms1 ms1 Y11 y11* plants gave rise to diploid, triploid, and tetraploid seedlings, and expressed green, yellow-green, and yellow phenotypes. Since apomixis does not occur in soybean, these results suggest that coenocytic pollen grains remain functional in *ms1 ms1* plants and participate in fertilization.

The embryo-endosperm relationship experiment was based on crosses of diploid (2n = 40) x tetraploid (4n = 80): *ms1 ms1 Y11 y11* x *Ms1 Ms1 Ms1 Ms1 Y11 Y11 Y11 Y11* genotypes. A total of 2007 crosses were made, and 32 F_1 seeds were obtained. Thirty-one F_1 seeds have been checked for chromosome number, and all seedlings were tetraploid. No triploid plants were found. Twenty-four F_1 plants were green, five F_1 yellow-green. The observed segregation ratio in F_1 was significantly different than the expected

1:1 ratio for plant color. From the results of this experiment, we conclude that only female gametophytes in which egg cells and polar nuclei underwent simultaneous duplication were able to participate in double fertilization and embryo development. We conclude that zygotes cannot represent independent ploidy level from primary endosperm, and the ratio of female to male genomes and endosperm balance number should be 2:1. Additional results will be obtained from F_2 populations tested for segregation of the *ms1* and *y11* alleles.

THE LEAF MRNA COMPLEMENT CHANGES DURING PHOTOPERIODIC FLOWER INDUCTION IN *SINAPIS ALBA*

F. CREMER, J. DOMMES, C. VAN DE WALLE, AND G. BERNIER

Department de Botanique, University of Liege, Bart Tilman, 4000 Liege, Belgium

Sinapis alba can be induced to flower by a single 22-h LD. Defoliation experiments suggested that the floral stimulus moves out of the leaves between 16 and 28 h after the start of the LD. In a preliminary attempt to correlate photoperiodic induction and modification of gene expression, we studied the changes of the leaf mRNA complement in plants induced to flower using *in vitro* translation of poly(A)+RNA and two-dimensional PAGE. The three youngest fully-expanded leaves were collected from induced plants 12, 16, 20, and 24 h after the start of the LD and simultaneously from control plants left in SD (8 h light - 16 h dark). Samples were also taken at the 4th and 8th h in SD.

The comparison of noninduced leaf mRNAs showed that about 10% of them present a diurnal change in quantity during the SD cycle. Numerous different diurnal cycles were observed.

The comparison of induced and noninduced leaf mRNAs showed that induction of flowering was accompanied by changes in some translatable mRNAs. Part of the mRNAs showing diurnal cycles in SD presented a modification of this cycle from the 12th or the 16th h of the LD; some other cycles were unaltered. Some mRNAs constant in SD showed quantitative modification from the 12th or the 16th h of the LD. Most changes disappear before the end of the LD.

In conclusion, the present results suggest that the leaf mRNA complement changes very early during the period of supplementary light of the inductive LD and in a very complex way. Unlike Warm and Lay-Yee *et al.*, we have been unable to show any qualitative change in mRNA complement during photoinduction of flowering.

THE USE OF GENETIC MOSAICS FOR STUDYING DEVELOPMENTAL MUTANTS

PHIL BECRAFT, NIC HARBERD, R. SCOTT POETHIG,
AND MICHAEL FREELING

Department of Genetics, University of California, Berkeley, CA 94720, USA

Mutants provide valuable tools for studying the genetic control of complex developmental processes. Mutants are available which affect nearly all aspects of plant development. We have been studying maize mutants which affect both floral and vegetative organs.

D8 and *Mp11* are tightly linked (likely allelic) dominant dwarfing mutations which are insensitive to applied gibberellin (GA). Because of their close phenotypic resemblance to recessive dwarf mutants which can be rescued by applied GA, we believe *D8* and *Mp11* may involve GA reception. These mutations have profound effects on floral morphology. Florets on normal ears possess only pistillate structures, while *D8* and *Mp11* ear florets also contain stamens. This 'anther ear' phenotype is due to derepression of stamen primordia which normally cease development as rudiments. Suppression of stamen development in ear florets is presumably controlled by the wild-type allele(s) of *D8* and *Mp11*. In the tassels of these mutants, unisexual staminate florets are formed as normal, but internode length is decreased.

Genetic mosaics were created to study cell automy and tissue specificity of these mutations. *D8 Lw/+ lw* and *Mp11 Lw/+ lw* heterozygous plants were subjected to X-irradiation to include chromosome breakage. Lemon-white *(lw)*, a recessive albino mutation located 4 map units distal to *D8* and *Mp11*, was used to mark chromosome loss. Loss of the chromosome arm carrying *D8 Lw* or *Mp11 Lw* produces - -/+ *lw*. Clones derived from such an event occur as sectors of albino tissue lacking *D8* or *Mp11* (-/+), in otherwise green *D8/+* or *Mp11/+* plants.

Ear florets derived from albino *(-/+)* subepidermal cell layers developed normal unisexual pistillate structures, while those derived from green *(D8/+* or *Mp11/+)* subepidermal tissues displayed the 'anther ear' phenotype, regardless of epidermal genotype. Similarly, tassel internodes elongated according to the genotype of the subepidermal tissues. Our results indicate that *D8* and *Mp11* act in subepidermal tissues, cell autonomously or with a short range of influence.

We are also interested in development of the ligule, a fringe of tissue separating the blade and sheath of maize leaves. The ligule is initiated by a periclinal cell division in the leaf epidermis. Since all lateral organs in plants are initiated by periclinal divisions, the control of ligule initiation likely involves processes basic to plant development. We performed a similar genetic mosaic analysis of *liguleless1* (*lg1*), a recessive mutation which eliminates both the ligule and auricle. Our results expressly demonstrate the power of this technique for studying plant development. The wild-type *Lg1* gene product acts autonomously in the epidermis for ligule formation, and in the subepidermal layers for auricle formation. Mosaic analysis is the only method that could demonstrate a spatial separation of functions for such a pleiotropic mutation.

IDENTIFICATION OF MEIOSIS-SPECIFIC NUCLEAR PROTEINS OF *LILIUM*

C. Daniel Riggs, Clare A. Hasenkampf, and Herbert Stern

C-016, Department of Biology, University of California,
San Diego La Jolla, CA 92093, USA

The process of meiosis is essential for the sexual reproduction of most higher eukaryotes. The two critical features which distinguish meiosis are the intimate association of homologous chromosomes (synapsis) and genetic recombination. Chromosome synapsis and the associated events of recombination are essential for the production of chromosomally-balanced gametes. Failure of either process for any chromosome pair very frequently results in aneuploidy. *Lilium* (lily) is the best higher eukaryotic system for studying meiosis because large quantities of highly synchronous cells, representing all stages of meiosis, can be collected for biochemical and molecular biological studies.

We have chosen to examine the protein profiles of premeiotic nuclei, nuclei from all stages of meiosis, as well as nuclei which have undergone meiosis and are proceeding through the microspore mitoses. Total cellular proteins from meiotic and nonmeiotic cell types have also been compared with the goal of identifying proteins which might play roles in homologous chromosome pairing, genetic recombination, and the unique first division which characterizes meiosis. To this end, we have identified nuclear proteins unique to early or late prophase, and substages thereof, and are now in the process of characterizing them. Of particular interest are several abundant meiotic proteins which could be components of the synaptonemal complex, a structure associated with the synapsis of homologous chromosomes. Antibodies have been raised against several of these proteins and progress toward their nuclear localization (by immunocytochemistry) and characterization will be presented.

DEVELOPMENTAL STAGING OF MAIZE MICROSPORES REVEALS A TRANSITION IN DEVELOPING MICROSPORE PROTEINS

Patricia Bedinger, Michael D. Edgerton, and Thomas Whelan

*Biology Department, University of North Carolina,
Chapel Hill, NC 27599-3280, USA*

A method for the preparation of developmentally staged microspores and young pollen from maize (*Zea mays*) has been designed. The resulting preparations retain a high degree of viability, and are of sufficient purity and quantity for biochemical analysis. Here, we present the analysis of steady-state protein and RNA populations associated with each stage. A major transition occurs during the stage that encompasses microspore mitosis, the asymmetric nuclear division producing the vegetative and generative nuclei. At least 14 differences between early and late stage proteins can be detected by one-dimensional SDS-PAGE, and many more can be detected by two-dimensional gel electrophoresis of proteins. *In vitro* translation of RNA isolated from staged microspores indicates that at least some of these differences could result from changes in microspore gene expression. In addition, we present the analysis of proteins associated with pollen walls at each of the developmental stages.

DIRECTED MOVEMENT OF LATEX PARTICLES IN THE GYNOECIA OF THREE SPECIES OF FLOWERING PLANTS

L. C. Sanders and E. M. Lord

*Department of Botany and Plant Sciences, University of California,
Riverside, CA 92521, USA*

The secretory matrix of the stylar-transmitting tract of angiosperms has been characterized as a nutrient medium for the growth of pollen tubes, acting to guide tubes to the ovules. When nonliving particles (latex beads) were artificially introduced onto the transmitting tracts of styles of *Hemerocallis flava*, *Raphanus raphanistrum*, and *Vicia faba*, they were translocated to the ovary at rates similar to those of pollen tubes. Direct observations were made on the movement of individual beads along the secretory epidermis in the style and ovary of *Vicia faba*. The transmitting tract may play an active role in extending tube tips to their destination in the ovary.

CHANGES IN PROTEIN AND mRNA COMPLEXITY OVER A TIME COURSE FOLLOWING A FLORAL INDUCTIVE EVENT

Martin J.G. Hughes and Brian R. Jordan

*Department of Biochemistry and Molecular Biology,
Institute of Horticultural Research, Littlehampton, Worthing Road,
Littlehampton, West Sussex BN17 6LP UK*

Proteins are being extracted from the SD plant *Pharbitis nil* over a period of time following either a 16-h inductive dark treatment, or a noninductive treatment of 16 h dark with a 30-min red break. The proteins from each time point are being analyzed using two-dimensional gel electrophoresis, and compared to detect changes associated with floral induction.

Poly(A)+mRNA is being extracted from the cotyledons at the same time points as the proteins. Following *in vitro* translation reactions, polypeptide products will be analyzed, again using two-dimensional gel electrophoresis. The results of these experiments may give an indication of the timing of biochemical events following floral induction, and thus suggest a time point from which mRNA or proteins may be extracted for further analysis. Results from these experiments will be presented and discussed.

STAMEN DEVELOPMENT IN THE TOMATO MUTANT *GREEN PISTILLATE*

Nicolas Rasmussen

*Department of Biological Sciences, Stanford University,
Stanford, CA 94305-5020, USA*

The tomato mutant green pistillate produced flowers showing strong apparent transformation of petals to sepals and stamens to carpels. The homeotic phenotype is completely recessive, and is represented in all flowers of the homozygote plants. The carpelloid stamens of mutant flowers take part in development of supernumary locules, which bear ovules, and no pollen is produced. Ovules in both the regular and supernumary locules can be fertilized with pollen from another flower, but all seeds of the homozygote flowers abort before the reach 1 mm in size. Histological preparations reveal that the carpelloid stamens follow a carpel-like developmental path from an early stage, and that supernumary locules produce ovules at approximately the same time as the regular locules of both mutant and wild type flowers. The production of supernumary ovules ectopically, on the external surface of carpels in mutant flowers, is suggestive of a possible inductive interaction between carpel surfaces in the normal differentiation of ovules. The results of SEM studies on very young floral apices, using the replica method of Michelle Williams, indicate no substantial difference in the epidermal cell pattern between mutant and wild type stamens at organ initiation. This evidence counts against the possibility that the biophysical situation at organ initiation by itself determines subsequent organ differentiation events in flower development.

CELLS FROM VEGETATIVE TOBACCO PLANTS HAVE THE CAPACITY TO FORM *DE NOVO* FLORAL SHOOTS

CARL N. MCDANIEL, KARLA A. SANGREY, AND DOROTHY E. JEGLA

Plant Science Group, Department of Biology,
Rensselaer Polytechnic Institute, Troy, NY 12180-3590 (C.N.M.)
and Kenyon College, Gambier, OH 43022, (K.A.S., D.E.J.), *USA*

In this study, we have characterized the developmental fate of *de novo* shoots formed in culture, and *in situ*, by stem tissues of *Nicotiana tabacum* cv Wisconsin 38 taken from plants of different ages and from different positions along the main axis. Only 0.1 to 0.3% of the shoots formed from cultured tissues of young vegetative plants were floral, while 1 to 2% of the shoots formed from the same region of plants that had flowered were floral. Cultured inflorescence tissue produced 76% floral shoots. Floral shoots were assumed to have arisen from florally-determined terminal and axillary buds; they produced several nodes and then a terminal flower. Some internode cells that were stimulated to undergo organogenesis *in situ* also produced few-noded floral shoots and, therefore, also were derived from florally determined cells or became florally determined as a result of their position on the main axis. There was a wide range, however, in the number of nodes produced by *de novo* shoots organized in the more basal internode positions. This can be interpreted to mean that floral determination in day neutral tobacco is not an all or none state. Our results indicate that the capacity of tobacco cells to form a flower is a visible indication of a developmental state that is controlled quantitatively, but not qualitatively, in temporal and positional terms.

MOLECULAR MARKERS FOR THE FLORAL PROGRAM

ELIEZER LIFSCHITZ

Department of Bioology, Technion-Israel, Institute of Technology, Haifa 32000, Israel

We surmise that upon induction, all meristematic cells fated to form the inflorescence acquire a "floral program" that distinguishes them from vegetative cells. The floral program maintans the floral "state" of all cells throughout differentiation and in all floral organs. It is activated by gene products that are responsible for the correct interpretation of the inductive cues, and it is a developmental prerequisite for the operation of the homeotic systems which regulate the timely appearance, structure, and function of a particular organ. Thus, our research is primarily concerned with attempts to describe the presumptive "floral program" by searching for genes and proteins that will mark flower-specific processes common to all floral organs, and all developmental stages.

Tomato and *Brassica oleracea* are our main experimental systems. Of particular importance is a recessive mutation, *anantha*, which allows for the formation of an ever-branching, cauliflower-like inflorescence in tomato. It is the source of the earliest possible flower-specific proteins and genes.

Using various separation techniques, we compared the appearance of more than 500 major proteins in leaves, flowers, and *anantha* meristems. Presumptive 'common denominator' proteins, as well as early- or late-appearing proteins were isolated. Antibodies were subsequently raised and used to screen expression cDNA libraries of *anantha* meristems and normal flowers.

The analysis of mRNA and the genomic organization of three flower-specific genes thus isolated will be described in detail. We will present a full description of the localization of mRNA and corresponding antigens in all floral organs and throughout development. This analysis reveals an heretofore undescribed differentiation of floral parenchymal cells using molecular markers.

In parallel with these studies, every antibody raised against the tomato flower is tested for recognition of a similar antigen in the cauliflower-broccoli system, which represents a temporal blockage of flower development in the meristematic and preflower stages respectively. Ab-P5 recognizes a glycoprotein that is rare in *anantha* meristems, but abundant in normal tomato flowers. It does recognize a glycoprotein specific to broccoli preflower, but not to cauliflower meristems, or to leaves of either species. AbP16 is specific to *anantha* meristem protein and recognizes an antigen specific to cauliflower meristems only.

Preliminary conclusions: (*i*) Some flower-specific genes operate throughout floral differentiation and in all floral organs. (*ii*) Proteins common to all floral organs are found also among "late" flower-specific genes. In fact, in many respects, petals and even sepals are demonstratively similar to stamens or ovaries and styles, rather than to leaves. (*iii*) Genes specific to the flowering state so operate in the *anantha* meristems. (*iv*) Parenchyma cells within a particular organ exhibit remarkable differences in expression of 'common denominator' genes.

EVIDENCE FOR TWO ACTIONS OF LIGHT IN THE PHOTOPERIODIC CONTROL OF FLORAL INDUCTION

PETER J. LUMSDEN

Lancashire Polytechnic, Preston, Lancashire PR1 2TQ, UK

Floral induction in SD plants requires exposure to a critical duration of darkness; this involves the interaction of a photoreceptor to distinguish between light and dark, and a timing system to measure the duration of darkness. The timing system appears to be based on a circadian rhythm, which is deduced from the changing flowering response to pulses of red (night-break) light given during an otherwise inductive dark period. This circadian flowering response rhythm is subject to phase control by light; short pulses of red light elicit changes in phase of the rhythm and are reversible by far-red light, while after continuous light of more than 6 h, the rhythm appears to be suspended, and the time of maximum inhibition of flowering by a night-break then occurs at a constant time of 8 h from the onset of darkness. Thus, under natural conditions, timing is coupled to the onset of darkness; induction occurs if the subsequent photoperiod does not impinge on the light-sensitive part of the rhythm.

The question arises, however, whether the inhibition of flowering is really due to a separate action of light, or is only a consequence of phase-shifting of the rhythm. The two responses (inhibition of flowering and phase shifting) were compared at the 6th and 8th h of darkness using brief red light treatments; they differed in their dose responses and, by using very short exposures, it was possible to achieve one response without the other. At the 6th h, a phase shift could be achieved without inhibition of flowering, while at the 8th h, flowering could be inhibited without a phase-shift, indicating two separate actions of light.

ISOLATION OF S-LOCUS GENES FROM
PETUNIA HYBRIDA

REED CLARK, JOHN OKULEY, PAMELA COLLINS AND THOMAS SIMS

Department of Molecular Genetics, Ohio State University, Columbus, OH 43210, USA

Our laboratory is investigating the molecular basis of gametophytic self-incompatibility in *Petunia hybrida*. As an initial step in investigating the molecular basis of gametophytic self-incompatibility, we have cloned putative S-locus genes from *Petunia hybrida* lines containing S_1, S_2, and S_3 alleles. We employed two different strategies for the isolation of S-locus genes. The first strategy was to isolate style-specific cDNA clones, then to characterize those clones for possible homology with the S_2 allele of *Nicotiana alata* (1). Next, we utilized direct screening of cDNA libraries using an oligonucleotide complementary to the *Nicotiana* S_2 cDNA sequence in a region known to be highly conserved among different S-alleles of *Nicotiana* and tomato.

The labeled oligonucleotide was hybridized to poly(A)+RNA from styles of $S_{1.2}$ or $S_{3.3}$ plants under conditions of low stringency. Under these conditions, we detected hybridization to an mRNA of about 950 nt present in styles, but undetectable in leaf or petal mRNA populations. The oligonucleotide was subsequently used to screen cDNA libraries constructed in λ ZAP. We have isolated a total of seven different cDNA clones from those screens. One clone, termed PHSB, has been partially characterized by DNA and RNA blots, restriction mapping and DNA sequencing. The PHSB clone hybridizes specifically to a 950 nt mRNA from $S_{1.2}$ and $S_{1.1}$ styles, but does not hybridize detectably to mRNAs from leaves, petals or anthers. Although the PHSB clone does not hybridize to stylar mRNA from the $S_{3.3}$ strain under moderately high stringency conditions, hybridization to $S_{3.3}$ mRNA is seen when low stringency conditions are employed. The mRNA detected in these experiments is highly prevalent in mature styles, but is present at much lower amounts in styles from immature floral buds. Preliminary DNA blot hybridizations indicate that the PHSB gene is single or low copy. Lastly, partial DNA sequence analysis indicates that the PHSB cDNA is homologous (about 70% nucleotide sequence conservation) to the S_2 cDNA of *Nicotiana alata* (1). Taken together, these data suggest that the cDNA clones we have isolated represent S-locus genes of *Petunia hybrida*. We will be utilizing these cloned genes as tools to decipher the molecular basis of self-incompatibility in *Petunia*.

1. ANDERSON MA, EC CORNISH, SL MAU, EG WILLIAMS, R HOGGART, A ATKINSON, I BONIG, B GREGO, R. SIMPSON, PJ ROCHE, JD HALEY, JD PENSCHOW, HD NIALL, GW TREGEAR, JP COCHLAN, RJ CRAWFORD, AE CLARKE 1986 Cloning of a cDNA for a stylar glycoprotein associated with the expression of self-incompatibility in *Nicotiana alata*. Nature (Lond) 321: 38-44

SELF-INCOMPATIBILITY IN *TRIFOLIUM PRATENSE* L (RED CLOVER) IDENTIFICATION AND CHARACTERIZATION OF S-LOCUS ASSOCIATED STYLAR PROTEINS

P. A. Eastman, S. R. Bowley, and B. E. Ellis

Department of Crop Science, (P.A.E., S.R.B.), and Department of Chemistry and Biochemistry (P.A.E., B. E. E.), University of Guelph, Guelph, Ontario N1G 2W1, Canada

Identification and characterization of S-locus-associated proteins isolated from red clover florets is being conducted to investigate the molecular basis of gametophytic self-incompatibility (SI) in a legume species. The protein extracts of four lines homozygous for different S-alleles and nine heterozygous F_1 crosses have been analyzed by SDS-PAGE, isoelectric focusing, and chromatofocusing. Proteins whose expression coincides with segregation of the different S-alleles have been identified on the basis of isoelectric point (pI) and partially purified. Physical characterization of the S_1-allele-associated protein has demonstrated that, *in vitro*, the native protein is a homodimer of approximately 130 kD with a pI of 6.0, and is N-glycosylated. Comparison of the characteristics of the red clover S-locus-associated proteins with those published for SI proteins from other species reveals features in common.

WHAT HAPPENS TO ORGANELLE DNA IN POLLEN?

J. L. Corriveau, L. J. Goff, and A. W. Coleman

Division of Biology and Medicine, Brown University, Providence, RI 02912, USA

In angiosperms, uniparental inheritance is correlated with the absence of plastid DNA in mature pollen, as detected both cytologically and by methods of molecular biology. Furthermore, cytological examination suggests that angiosperm pollen generally lacks mitochondrial DNA as well. This is compatible with what is known genetically of mitochondrial inheritance. Molecular biology examination of the situation is currently underway.

Acknowledgment--Supported by USDA grant 87-CRCR-1-2534.

MOLECULAR ANALYSIS OF GENE EXPRESSION DURING POLLEN DEVELOPMENT IN *OENOTHERA ORGANENSIS*

Sherri M. Brown and Martha L. Crouch

Biology Department, Indiana University, Bloomington, IN 47405, USA

Present address: Monsanto Company, 700 Chesterfield Village Parkway, Chesterfield, MO 63198, USA (S.M.B.)

We are analyzing the expression of several genes in the mature pollen grain of *Oenothera organensis*, the evening primrose. cDNA clones were isolated by differentially screening a library constructed from mRNA present in mature *Oenothera* pollen. Northern analysis shows that genes expressed in pollen have a variety of temporal and spatial patterns of mRNA accumulation. One class of cDNAs detects mRNAs only in pollen. mRNAs in this class accumulate in a variety of patterns throughout pollen development, but none of the transcripts are detectable before the mitotic division that produces the vegetative and generative cells. A second class of cDNAs detects mRNAs in other parts of the plant, as well as in pollen. These mRNAs, unlike the first class, are detectable at all gametophytic stages tested. Similarly, mRNAs encoding actin, tubulin, and histone are detectable in all parts of the plant, as well as in all stages of pollen development. Genomic Southern blots show that the mRNAs present primarily in pollen are from gene families of varying sizes.

Expression of one gene family (P1/P2) was examined in detail. mRNAs from this family are abundant in mature *Oenothera* pollen (0.013% of total RNA), and are detectable in the pollen from several unrelated plant species. Antiserum against part of the protein-coding region of one of the cDNAs was used to characterize the protein products of this gene family. The P1/P2 gene family produces a family of polypeptides (40 to 45 kD) that is present in pollen at late stages of development, in mature pollen, and in pollen tubes. The nucleotide sequence of a cDNA clone representing nearly the entire length of the mRNA was determined and the amino acid sequence deduced from the nucleotide sequence. Both nucleotide and amino acid sequences show striking similarity to the published sequences for cDNAs encoding the enzyme polygalacturonase, suggesting that the P1/P2 gene family may function in breaking down cell walls during the pollination process.

THE EFFECTS OF CHEMICAL HYBRIDIZING AGENTS SC-1058 AND SC-1271 ON THE ULTRASTRUCTURE OF DEVELOPING WHEAT ANTHERS (*TRITICUM AESTIVUM* L. VAR. *YECORA ROJO*).

JOHN W. CROSS, THOMAS PATTERSON, JEFFREY LABOVITZ, EDUARDO ALMEIDA AND PATRICIA J. SCHULZ

Sogetal Inc., 3872 Bay Center Place, Hayward, CA 94545 (J.W.C., T.P., J.L.); *and Department of Biology, University of San Francisco, San Francisco, CA 94117* (E.A.; P.J.S.), *USA*

The chemical hybridizing agents (CHA) SC-1058 and SC-1271 induce the abortion of developing pollen grains without interfering with ovule development or female fertility. Application of these agents to wheat allows production of hybrid wheat without tedious hand emasculation. We have studied the morphology of wheat anthers treated with these CHA compounds to determine their mode of action. Potted wheat plants were sprayed with the compounds, or with a blank spray formulation, at the 2-cm spike stage. We used sufficient compound to cause complete male sterility. The anthers were then fixed and examined by light and electron microscopy.

At the time of spraying, the pollen mother cells were in meiotic prophase. No effect on meiosis was seen in stained anther squashes under the light microscope. When compared with the control material, chemically-treated plants did show a significant slowing of development, beginning during meiosis. After meiosis, there was an abnormal development of both the tapetal cells and the microspores.

Tapetal Cells. These cells became prematurely vacuolate. They failed to flatten tangentially (as occurs normally during the enlargement of the microspores), and they became electron-dense. The Ubisch bodies and the orbicular wall (both composed of sporopollenin) did not form correctly. Electron-dense material was deposited in small vacuoles, and there was abnormal plastid development.

Microspores. Young, uninucleate microspores also became prematurely vacuolate. The exine wall (composed of sporopollenin) was malformed, and was thinner than in the control. The tonoplast membrane ruptured and the cytoplasm of the microspores broke down, causing the microspores to collapse at the vacuolate stage of development. Microspore mitosis may or may not occur before death ensues.

Acetolysis of whole anthers and anther sections (a treatment that destroys all structural components other than sporopollenin) reveals that the sporopollenin-containing components of CHA-treated anthers are thinner and more flexible than in the control. However, we believe that this effect was secondary.

Our observations indicate that these chemicals first induce changes in tapetal cell function which adversely affect the development of the microspores. We hypothesize that these chemicals interfere with the secretion of nutrients or structural building-blocks for the developing microspores.

USE OF T-DNA TAGGING IN *ARABIDOPSIS* TO GENERATE MUTANTS IN FLOWER DEVELOPMENT

K. A. FELDMANN

Plant Sciences, Central Research and Development Department, E. I. DuPont, Wilmington, DE 19898, USA

Screening of 400 transformed lines of *Arabidopsis thaliana*, generated by infection of seeds with *Agrobacterium*, has resulted in 18 mutant lines that segregate in a Mendelian manner for defects in fertility. The *Agrobacterium* strain (C58C1rif, 3850:1003 plasmid) contains a dominant, kanamycin resistance marker between the T-DNA borders. The transformed lines were screened by growing 30 to 40 seeds to maturity under growth chamber conditions. The phenotype of the infertile plants generally includes reduced seed set and thus short siliques, long internodes, and plants that are typically taller than wild-type plants. Microscopic examination of a few of these mutants has revealed that one line fails to produce viable pollen. In another line, filaments fail to elongate such that pollen is not shed on the stigmatic surface. Yet another line exhibits perturbations of the stigmatic surface and, finally, in low lines, seeds fail to form in the basipetal end of the silique. Genetic characterization reveals that all mutations examined are recessive to the wild-type alleles. These mutants are not caused by somaclonal variation, because they were generated in a nontissue culture manner. These mutants may have arisen from a T-DNA insert, or a spontaneous mutation. Cosegregation tests to ascertain whether the kanamycin resistance marker and the mutant allele are linked are in progress. Genetic characterization of at least two unrelated mutants has shown that the kanamycin resistance marker and the mutant allele cosegregate; consistent with the hypothesis that a T-DNA insert has knocked out the function of a gene which, in the homozygous condition, causes an altered phenotype. A decription of the various flower mutants, along with accumulated cosegregation data, will be presented.

STAGE-SPECIFIC CHANGES IN PROTEIN COMPLEMENT ACCOMPANY SPIKELET DEVELOPMENT IN THE MAIZE EAR

Teresa S. Findlay and Alan R. Orr

*University of Northern Iowa,
Cedar Falls, IA 50614, USA*

Analyses of *Zea mays* strains that exhibit mutations at specific stages in the organogenesis of the ear and tassel indicate that the initiation/development of primordia is genetically regulated at each successive morphological stage in the formation of floral organs. A long-term goal of our laboratory is to investigate the relationship between the morphological events and the underlying biochemical changes in maize inflorescence development. The soluble proteins of three successive developmental stages (spikelet primordia, glume primordia, and floret primordia) of the ear were analyzed by high resolution two-dimensional PAGE. The results showed a pattern of qualitative changes in the protein complement that are coincident with the morphological events. Using two-dimensional PAGE, one of us previously demonstrated a stage-associated pattern of protein change specific to earlier stages (apical meristem and branch primordia) of ear and tassel development. Taken together, these observations indicate that the control of maize inflorescence morphogenesis is, in part, sequentially imposed at successive developmental stages.

ROLE OF ETHYLENE IN POSTPOLLINATION EVENTS

A. D. Stead, S. Han, and M. S. Reid

*University of London, London, UK (A.D.S.), University of Massachusetts,
Boston, MA 02116 (S. H.), University of California,
Davis, CA 95616 (M.S.R.), USA*

In many species, the consequences of successful pollination can be seen within a few hours of pollination. Events such as wilting, abscission, or color changes of the corolla, have all been shown to occur rapidly after pollination and before fertilization could occur. These changes invariably have been linked to changes in either the production of, or sensitivity to, ethylene by the floral tissues.

The events following pollination in *Digitalis*, *Petunia*, *Brodiaea*, and *Lupinus*, include pollination-induced corolla abscission, corolla wilting, and changes in pigmentation. In all of these species, increased ethylene production has been detected from the stigma and/or style within a few minutes of pollination, and in each the response can be shown to occur if the tissue is exposed to exogenous ethylene. Inhibitors of ethylene synthesis (AVG; Cobalt ions), or action (STS), will prevent or delay these pollination-induced changes. In *Digitalis* at least, the response is quantitatively related to pollen loading, greater amounts of pollen eliciting greater ethylene production and faster corolla abscission. The total amount of ethylene evolved prior to abscission is, however, constant--suggesting that pollination controls the rate of synthesis of a predetermined amount of ethylene. In *Brodiaea*, unpollinated flowers produce little or no ethylene during senescence, but pollination results in wilting associated with induction of ethylene biosynthesis.

In *Petunia* and *Lupinus*, inhibitors such as cycloheximide prevent the entylene-induced changes, indicating that *de novo* protein systhesis is required for the response. Furthermore, at least in *Petunia*, actinomycin D prevents many of the changes that occur as a result of exposure to exogenous ethylene, indicating that ethylene, produced as a result of pollination, induces changes in gene expression in this species.

DEVELOPMENT OF THE DIMORPHIC ANTHERS IN *COLLOMIA GRANDIFLORA*; EVIDENCE FOR HETEROCHRONY IN THE EVOLUTION OF THE CLEISTOGAMOUS ANTHER

E. M. LORD, K. J. ECKARD, AND W. CRONE

Department of Botany and Plant Sciences, University of California, Riverside, CA 92521, USA

A modification characterizing all cleistogamous species is reduction in anther size of the CL (cleistogamous or closed) flower. In *Collomia grandiflora* the CL anther, in addition to being smaller, has only two locules; the CH (chasmogamous or open) flower anther has four locules. As a consequence, there is a modification in CL anther shape. From initially similar primordia, a divergence in histology between the two anther forms appears at archesporial cell differentiation when only two locules are established in the CL anther. The process of form divergence in the two anther types is examined in this study using histological, allometric, and three-dimensional computer graphic techniques. Allometric data from SEM images demonstrate the equivalence of primordial shapes at anther inception and divergence just prior to archesporial cell division, which signals the onset of sporogenous cell proliferation. Reconstructions of the anthers at archesporial cell division stage revealed differences in external and internal form and size, features unrelated to locule number. Fewer initial archesporial cells and a shorter duration of sporogenous cell proliferation in the CL anther correlates with a smaller anther with 1/10th the number of pollen grains at maturity. The CL anther shows less cell division activity from the time of archesporial cell division and no trace of the intercalary growth which appears during meiosis in the CH anther. The divergent CL anther size and form may be attributed to an earlier onset of abaxial locule differentiation in a smaller primordium which may itself preclude adaxial locule initiation. Heterochrony, or alteration in developmental timing, is proposed as the mode of evolution of the CL from the ancestral CH form.

MERISTEM CULTURE TO ASSAY FLORAL DETERMINATION IN TOBACCO

S. M. E. SMITH AND C. N. MCDANIEL

Plant Science Group, Department of Biology, Rensselaer Polytechnic Institute, Troy, NY 12180-3590, USA

Floral determination of the shoot apical meristem is a developmental state which can be assayed by placing the putatively determined meristem into isolation and observing its behavior. A meristem is considered florally determined if it forms a flower immediately, or after producing several nodes. This state has been assessed in tobacco by rooting terminal or lateral buds; buds included the meristem proper, plus 8 to 12 leaves and leaf primordia (1, 2). The effect of these young leaves on the timing and stability of floral determination in a meristem is unknown. Meristems of tobacco with as few as three leaf primordia can be cultured *in vitro* on hormoneless medium and will grow into normal mature plants (3). Employing this culture procedure, the timing and stability of floral determination is being assayed in meristems of *Nicotiana tabacum* cv Wisconsin 38, Hicks and Hicks Maryland Mammoth. We are particularly interested in whether the stability of the floral state in the meristem of the SD variety Hicks Maryland Mammoth differs from that observed in meristems of the day neutral varieties Hicks and Wisconsin 38.

1. MCDANIEL CN 1978 Determination for growth pattern in axillary buds of *Nicotiana tabacum* L. Dev Biol 66: 250-255
2. SINGER SR, MCDANIEL CN 1986 Floral determination in the terminal and axillary buds of *Nicotiana tabacum* L. Dev Biol 118: 587
3. SMITH, R. H., AND T. MURASHIGE 1970 *In vitro* development of isolated shoot apical meristems of angiosperms. Am J Bot 57: 562-568

FLORAL DEVELOPMENT IN *ARABIDOPSIS THALIANA*: A COMPARISON OF THE WILD TYPE AND THE HOMEOTIC PISTILLATA MUTANT

JEFFREY P. HILL AND ELIZABETH M. LORD

Department of Botany and Plant Sciences, University of California, Riverside, CA 92521, U.S.A.

Homeosis is sometimes defined as the replacement of one member of a meristic series by another member normally formed in a different position. We examined the developmental basis in a case of the pistillata floral mutant of *Arabidopsis thaliana* (Brassicaceae). Pistillata (PI) has petals replaced by sepal-like organs ("petals"), is male sterile, and has abnormal gynoecial development. Normal sepal development in wild type (WT) and pistillata flowers is indistinguishable in terms of initiation events, anatomical and morphological development, and allometric growth. Wild type petals and PI "petals" are initially ontogenetically similar in these same respects. The first observable difference between WT and PI floral development is abnormal patterns of cell division in PI at stamen initiation. Anomalous patterns of cell division and ontogenetic variability in mutant flowers contribute to a wide range of final morphologies. Tissues which normally form the androecium appear to be congenitally fused to the gynoecium to various extents, and differentiate gynoecial tissues. These structures may be filamentous in form and capped by stigmatic cells, or may consist of more substantial tissue masses which differentiate as unfused capels bearing naked ovules. Form divergence between WT petals and PI "petals" becomes evident when these organs reach 90 μm in length, after androecial developmental divergence has occurred. Pistillata does not have petals replaced by sepals; PI "petals" are intermediate in form between WT sepals and petals due to the developmental switching of petal primordia into the ontogenetic pattern characteristic of WT sepals after petal primordia are initiated. PI "petals" may still be considered homeotic if Bateson's (1894) original broad definition of the term is used. We attribute the developmental basis for PI "petals" to heterochrony, which is defined as a change in the timing of an important developmental event in the mutant relative to the WT. The homeotic variation seen in PI "petals" provides insights into the developmental genetic control of organ determination and demonstrates the morphogenetic lability of plant primordia.

POLLEN TUBE GROWTH AS A FUNCTION OF DONOR AND RECIPIENT RELATEDNESS

JOHN NASON AND NORMAN ELLSTRAND

Department of Botany and Plant Science, University of California, Riverside, CA 92521, USA

Self-incompatibility systems (sporophytic and gametophytic) have evolved through selection against the production of inbred progeny through selfing or mating among close relatives. In populations having numerous S-alleles with complex dominance relationships, however, a large proportion of crosses between close relatives may yet be compatible. It is possible that in self-incompatible populations, mechanisms may have also evolved which act in compatible crosses among relatives to alter patterns of pollen tube growth and, consequently, the production of inbred progeny.

This study examines compatible pollen tube growth as a function of the genetic relatedness of mates in wild *Raphanus sativus* (Brassicaceae), an annual with a sporophytic self-incompatibility system. For each gynoecium, we scored the number of germinated pollen grains and the number of pollen tubes reaching 0.0, 1.5, 3.0, and 4.5 mm below the base of the stigma in compatible crosses between individuals representing five different levels of genetic relatedness (Wright's coefficient of inbreeding, f, equal to 0.0, 0.0315, 0.125, 0.156, and 0.25 by pedigree). The effect of relatedness on the number of germinated pollen grains was not found to be significant. In order to account for non-independence in the number of pollen tubes scored at successive distances in a given style, a repeated measures multivariate analysis of variance was used to examine the effect of relatedness on pollen tube growth. The primary results indicate that there is a significant interaction (variation in regression slope) between relatedness and distances reached by pollen tubes, but the effect of relatedness on pollen tube number is not significant when averaged across the different distances. With the removal of one outlying value from the analysis, however, not only is the interaction between relatedness and distance significant (Wilks' λ $Pr > F = 0.045$), as above, but when averaged over the different distances below the stigma, the effect of relatedness on pollen tube number is also found to be significant ($Pr > F = 0.031$).

ACQUISITION OF COMPETENCE FOR FLORAL DEVELOPMENT IN *NICOTIANA* BUDS

SUSAN R. SINGER, COLE H. HANNON, SARAH C. HUBER

Department of Biology, Carleton College, Northfield, MN 55057, USA

Floral determination in dayneutral *Nicotiana tabacum* cv Wisconsin 38 occurs in response to a pervasive signal (2). Acquisition of this new developmental state is precisely regulated, and the terminal bud produces about four nodes after the time of floral determination (1). A major question arising from this work is whether a developmental signal reaches a critical concentration in terminal buds at the time of floral determination, or whether terminal buds gain competence to respond to an existing signal at that time. Preliminary data indicate that the latter hypothesis is true for *N. tabacum* cv Wisconsin 38, while the former is true for LD *N. sylvestris*.

Young seedlings of both *N. tabacum* and *N. sylvestris* produce equivalent numbers of nodes when grafted into the same apical position on flowering *N. tabacum* stocks. This number of nodes is less than that produced by seed-derived plants, but significantly more than the four nodes produced by florally determined buds. Grafting both seedling genotypes to flowering *N. sylvestris* stocks results in a further reduction in the number of nodes produced before terminal flower initiation, indicating that signal strength is critical in floral determination. Experiments varying stock age indicate that signal strength is constant over time for *N. tabacum*, but increases prior to flowering in *N. sylvestris*. Grafting *N. tabacum* terminal buds of different ages to apical positions on flowering *N. tabacum* stocks indicates that these buds may become more competent to respond to a developmental signal near the time of floral determination.

Acknowledgment--Supported by USDA grant 87-CRCR-1-2554, and a William and Flora Hewlett Research Corporation grant.

1. SINGER SR, McDANIEL CN 1986 Floral determination in the terminal and axillary buds of *Nicotiana tabacum* L. Dev Biol 118: 587
2. SINGER SR, McDANIEL CN 1987 Proc Natl Acad Sci USA 84: 2790

ISOLATION OF cDNAs FROM SHOOT TIPS OF MORNING GLORY SEEDLINGS INDUCED TO FLOWER

CAROLE L. BASSETT, TERESA GRUBER, AND DEBRA MOHNEN

*USDA-ARS, Russell Research Center, Athens, GA 30613 (C.L.B.);
Complex Carbohydrate Research Center, University of Georgia,
Athens, GA 30602 (T.G., D.M.), USA*

The Japanese morning glory, *Ipomoea [Pharbitis] nil*, can be induced to flower by exposure of young seedlings to a 16-h dark period. During evocation, the shoot tips respond to an unidentified signal(s) by initiating cellular and biochemical changes which ultimately result in the conversion of the vegetative growth pattern to a predominantly floral one. It is generally agreed that at some point this transition must involve alterations in gene expression in apical and axillary meristems. To isolate sequences whose regulation might be altered during this transition stage, we isolated poly(A)+RNA from shoot tips of 11-d-old seedlings which had been photoperiodically induced at day 4 and constructed a cDNA library in λ gt10. The resulting plaques were screened several times with ^{32}P-labeled cDNA synthesized from poly(A)+RNAs isolated from both induced and vegetative (control) shoot tips. Three classes of sequences have so far been identified: ind+ (expression predominant in induced shoot tips), veg+ (expression predominant in uninduced shoot tips), and neutral (expression not altered between induced and uninduced plants).

ISOLATION OF GENES EXPRESSED IN THE FLORAL MERISTEM

JUNE MEDFORD, J. SCOTT ELEMER, AND HARRY KLEE

*Plant Molecular Biology Group, Monsanto Company,
700 Chesterfield Village Parkway, St. Louis, MO 63198, USA*

The conversion of the shoot apical meristem from vegetative to reproductive development represents a precise conversion from indeterminate to determinate development in plants. Since the shoot apical meristem consists of less than 50 µg of tissue and is tightly surrounded by numerous leaves, it has been inaccessible to molecular analysis. We took advantage of the fortuitous amplification of meristematic material in *Brassica oleracea* (cauliflower and broccoli) and constructed cDNA libraries enriched for mRNA expressed in meristematic regions. Clones that are preferentially expressed in the meristems were identified using differential plaque hybridization. Promising clones were further characterized by Northern and Southern hybridization. For one initial *Brassica* clone, the corresponding genomic clone from *Arabidopsis thaliana* has been isolated. The transcriptional promoter was defined by primer extension and fused to the reporter gene β-glucuronidase (GUS) and the GUS gene fusion was introduced into a plant transformation vector. Analysis of transgenic tobacco plants using a GUS histochemical assay shows that the promoter preferentially directs transcription in differentiating reproductive tissues in the shoot apex. Analysis of other clones showing greater specificity for the shoot apical meristem is currently in progress. Since many of the processes related to differentiation and organ formation are localized in the vegetative and reproductive meristems, this approach will provide fundamental molecular details of events in the floral meristem.

A SEQUENTIAL STUDY BY SCANNING ELECTRON MICROSCOPY OF A DEVELOPING FLORAL APEX

MICHELLE H. WILLIAMS AND PAUL B. GREEN

Department of Biological Sciences, Stanford University, Stanford, CA 94305-5020, USA

Studies of organogenesis at plant meristems have been hampered by the inability to combine sequential observations with high resolution views of cellular dynamics. A two-step replica technique has been successfully developed for the sequential study of epidermal cell patterns of apical meristems by scanning electron microscopy. This technique is nondestructive and allows observations of the same tissue during its development. In brief, the method involves spreading a dental impression material on the floral apex which is left to set for 5 to 10 minutes. The negative moulds are then peeled from the plant and filled with epoxy resin, which is polymerized. The positive resin replicas are peeled from the negative moulds and mounted on scanning electron microscope stubs, sputter coated, and observed. Here, results are presented which demonstrate the advantages of such a method when applied to a floral apex. Patterns of cell division and expansion are clearly visible and can be related to organogenesis.

SALICYLIC ACID AND THERMOGENICITY IN SOME PLANT SPECIES DURING ANTHESIS

HANNA SKUBATCH, ILYA RASKIN, BASTIAAN J.D. MEEUSE,
AND ARNOLD J. BENDICH

*University of Washington,
Seattle, WA 98195, USA*

The level of salicylic acid (SA) was determined in the flowers and inflorescences of a number of thermogenic and nonthermogenic plant species in an attempt to understand the role of SA in heat production during floral development. SA was found in eight out of nine species investigated at levels up to 9 µg/g fresh weight.

In various inflorescence organs of four thermogenic species, *Amorphophallus campanulatus, Arum italicum, Arum dioscoridis,* and *Philodendron selloum*, SA was present at a high level on the first day of flowering (D-day), and in low concentration before and after D-day. In the male flowers of *Philodendron selloum*, SA level was 4-6 µg/g fresh weight during the first of two thermogenic phases, and the level dropped to 1 µg/g fresh weight during the second thermogenic phase, despite comparable heat production for the two phases.

SA was undetectable in the thermogenic flowers of *Victoria cruziana* (a water lily), indicating that SA is not required for heat production in all thermogenic plants.

The changes in the level of salicylic acid do not appear to be associated with the opening of flowers *per se*, as nonthermogenic flowers of *Passiflora caerulea, Tradescantia virginiana,* and *Oenothera biennis*, which have a short blooming period (less than 1 d), contained low and constant levels of SA throughout the anthesis period.

PHYSIOLOGY OF DEVELOPMENT AND SENESCENCE OF CAPITULUM IN *CHRYSAMTHEMUM*

P. Pardha Saradhi and H. Y. Mohan Ram

Centre for Biosciences, Jamia Millia Islamia, New Delhi-110 025 (P.P.S.) *and Department of Botany, University of Delhi, Delhi-110 007,* (H.Y.M.R.) *INDIA*

Investigations were carried out to understand the physiology of development and senescence of capitulae in *Chrysanthemum morifolium* cv Jyotsna (Asteraceae). Uniform young capitula, measuring 10 mm in diameter, were tagged on day 0 and sufficient number of capitulae were harvested at regular intervals. Bracts, ray florets, disc florets, and thalamus were excised and their fresh weight, dry weight, sugars (reducing and total), starch and protein contents (expressed as percent increase per gram dry weight) were determined. Activities of acid phosphatase(s) (orthophosphoric-monoester phosphohybrolase; EC 3.1.3.2), acid invertase (β-fructofuranosidase; EC 3.2.1.26), $\propto$-amylase (1,4-$\propto$-D-gluconglucano-hydrolase; EC 3.2.1.1), and peroxidase(s) (Donor: hydrogen peroxide oxidoreductase; EC 1.11.1.7) were estimated (on per g dry weight basis). The capitula showed maximal increase in diameter, fresh weight and dry weight by day 28. As expected, fresh weight and dry weight in the case of both ray florets and disc florets increased til day 28. However, in bracts, increment was restricted after day 20, while thalamus showed increment even after day 40. The maximum protein content was recorded on day 4 (2.5%), 22 (6.4%), 14 (7.6%), and 8 (6.2%) in bracts, thalamus, ray florets, and disc florets, respectively. The content of reducing sugars/total sugars in ray florets and disc florets increased until day 26 to 39.5/45.6 and 27.9/30.1, respectively. Activity of acid invertase was maximum in ray florets (ca 2,400 units) followed by disc florets (ca 850 units), thalamus (ca 515 units), and bracts (ca 101 units). The maximum acid phosphatase activity was approximately 510 units in ray florets, followed by 340 units in disc florets, 150 units in thalamus, and 101 units in bracts. Activity of these two enzymes in ray florets and disc florets coincided with the optimal growth period of the capitulum. No appreciable difference was observed in the activity of $\propto$-amylase in ray florets, disc florets and thalamus. The highest starch content recorded was 7.9 and 10.9 in ray florets and disc florets, respectively. All the above-mentioned parameters showed a decline over subsequent days. However, activity of peroxidase(s) was low during development and showed a sharp increase after day 28. The activity was highest in bracts (ca 6,000 units), followed by disc florets (ca 485 units), ray florets (ca 135 units) and thalamus (ca 6.4 units). The results are discussed with reference to the role of sugars, starch, protein, and enzyme activities in development and senescence.

FLORAL DEVELOPMENT AND HOMEOSIS IN *BEGONIA SEMPERFLORENS-CULTORUM* 'CINDERELLA'

Naida Lehmann and Rolf Sattler

McGill University, Montreal, Quebec, CANADA

The male flower in single-flowered begonias often has two broad outer tepals (sepals) and two smaller inner tepals (petals) in a more or less decussate arrangement, and an androecium consisting of numerous stamens. The male flowers of *Begonia semperflorens-cultorum* 'Cinderella' have tepaloid appendages in positions that stamens occupy in the single-flowered plants. This expression of tepal features where stamens are expected is an example of homeosis. Homeosis is a process whereby features of one structure are formed in the position of another structure. Homeosis can either completely or partially replace one structure with another. In the 'Cinderella' cultivar studied, the appendages in the androecial region vary in the expression of tepal features. Some appendages are much line the outer tepals, some are like that of the inner tepals, while others are more like stamens that have tepal tissue and anther locules.

The early stages of floral development in 'Cinderella' are similar to the early stages of development in a single-flowered cultivar that was also studied, while the latter stages of development diverge remarkably. The apex becomes laterally elongate and domed soon after the two outer tepals are initiated. The inner tepals are initiated slightly off from being in alternation with the two outer tepals. The first primordia that will develop into the tepaloid appendages are initiated adjacent to the inner tepals. Subsequently, one to three primordia arise opposite each of the outer tepals. Next, primordia arise in positions somewhat alternate to the existing primordia. From the beginning, however, phyllotaxy is irregular and continues to be irregular throughout development, so that the initiation of primordia cannot be described as being formed in any clear pattern. The first primordia that will form tepaloid appendages are small and round at the time of initiation, similar in appearance to those primordia that become stamens in the first-flowered plants. After the initiation of the first several primordia, the surface of the apex becomes irregularly shaped, and then it frequently divides into two or three apices. The primordia that form on these newly formed apices are flattened from the time of inception and look like those primordia that eventually form the outer tepals. Primordia continue to be initiated on these apices until the flower is mature and opens.

GYNOECIAL DEVELOPMENT IN *LILIUM LONGIFLORUM*: A KINEMATIC ANALYSIS

W. Crone and E. M. Lord

Department of Botany & Plant Sciences, University of California, Riverside CA 92521, USA

While many accounts describe the initiation and histogenesis of the gynoecium of flowering plants, they lack a kinematic sense of growth that cannot be reconstructed from sections alone. This study used surface marking experiments on *Lilium longiflorum* (Liliaceae) to examine the post-initiation growth of gynoecia *in situ*. Plants were equilibrated in a growth chamber for derivation of a normal growth curve for the entire bud and for an allometric plot of gynoecial length versus bud length (n=73 buds). Marking experiments represented two years of work, with 38 marked buds grown in the greenhouse in the first year and 32 marked buds grown in the growth chamber for the current experiments. The findings were consistent for the two groups.

For marking experiments, a segment of the perianth and intervening anthers were removed to expose the gynoecium, which was then marked longitudinally with an approximately equidistant row of dots of activated charcoal mixed with water. Precautions were taken to minimize desiccation and wounding effects. These marked gynoecia were then photographed every 24 h until growth ceased. The negatives were enlarged and the distance between recognizable features on consecutive marks measured by a digitizing bit pad. Any change in the distance between two points, when compared to the original length, gave a measure of the mean relative rate of elongation for that segment of the gynoecium; this was plotted against the distance from the base of the gynoecium. Gynoecia too small to be marked were serially sectioned to document patterns of cell division in the whole gynoecium.

Growth was shown to be exponential throughout the range measured (buds 3.5-170+ mm), with a relative bud growth of 0.0939/day. An allometric constant of 1.253 between bud length and gynoecial length was also demonstrated, giving a gynoecial relative growth rate of 0.118/day.

Marking experiments showed that the gynoecium grew in three phases. Phase one included gynoecia up to 7 mm in length (ca 10 mm bud). This was a period of fluctuating positions of maximal growth. The gynoecium grew as a unit, with no distinctions noticeable among the stigma, style, or ovary at this point. Two peaks in cell division activity were apparent in a sectioned 1.5 mm gynoecium. Phase two of gynoecial growth, for gynoecia between 7 and 14 mm (bud size of 10 to 20 mm), was a period of steady and uniform growth along the gynoecium. In phase three, for gynoecia longer than 14 mm, most of the growth occurred in the style, with the ovary and stigma growing at a more moderate rate. The style showed predominantly basal growth initially in this phase.

INDEX

ABA, 117, 119
Abortion, 97, 99, 101, 103, 114, 122, 124, 187
ACC, 100
Acid invertase, 43, 44, 101, 198
Actin, 111
Actinomycin D, 190
Agrobacterium tumefaciens, 130, 188
Alcohol dehydrogenase, 111
Amorphophallus campanulatus, 197
Aniline blue, 150, 166
Anther, 87, 89-93, 97, 108, 109, 115, 118, 121, 122, 124, 125, 128, 130, 132, 134, 166, 187, 190, 191
Anthesis, 98, 99, 103, 109, 111
Antirrhinum, 143
Apical dominance, 98
Apomixis, 175
Apple, 32-34, 39
Arabidopsis thaliana, 106, 134, 188, 196
  agamous mutant, 106, 107
  apetala-2, 106, 107
  apetala-3, 106
  double-mutant, 106
  pistillata 106, 107, 192
Arum dioscoridis, 197
Arum italicum, 197
Auxin, 130
AVG, 100, 190
Axiality shift, 62, 66, 69, 71

Barley, (see *Hordeum vulgare*) 2-6
Begonia semperflorens-cultorum, 199
Benzylaminopurine, 99, 101
Beta-galactosidase, 111
Beta-glucuronidase, 130, 131, 196
Beta-microglobulin, 162
Biophysics, 58, 66-68, 181
Bipolaris maydis, 122, 123, 125
Blasticidin S, 140

Brassica, 44, 140, 142, 143, 152, 153, 156-158, 161-163, 196
Brassica campestris, 44, 146, 148, 149, 152, 153, 161, 168, 170
Brassica oleracea, 136, 137, 139, 140-142, 146-148, 152, 153, 160, 168-170, 195
Brassica napus, 114, 117, 148
Broccoli, (see *Brassica*) 195
Brodiaea, 190

Cab gene, 7
Callose, 166, 167, 171
Calmodulin, 7
Carpel, 71, 106, 117, 118, 180, 181
Cauliflower, (see *Brassica*) 195
CCC, 99, 117, 119
Cell autonomy, 177
Cell division, 45, 63, 84, 89, 91, 93, 101, 196, 200
Cell wall, 108, 115, 138, 139, 167, 171
Cellulose
  microfibrils, 60
  reinforcement pattern, 61, 63
CEPA, 100
Chromatin, 47, 48
  structure, 47
  decondensation, 48, 147
Chromosome walking, 107
Chrysanthemum morifolium, 198
Circadian rhythm, 1-7, 183
Cleistogamy, 190
Collomia grandiflora, 190
Computer graphics, 190
Conifer, 29
Corn, 108-112, 121-125, 134, 174, 177, 179, 189
  mutants, 177
Correlative influences, 98, 99
Corolla abscission, 100, 190

Crassula argentea, 58
Cycloheximide, 140, 143, 190
Cytokinin, 45, 77, 80, 99, 101, 102, 117, 131, 136
Cytoskeleton, 60, 62
Digitalis, 190
2,4-dinitrophenol, 44

Easter lily (see *Lilium longiflorum*)
Echeveria, 64
E. coli, 123
Electron transport chain, 125
Endosperm, 166, 175
Environmental control, 95
    temperature, 95, 97, 106, 114, 117, 118
    light, 95, 96, 183
    daylength, 1-3, 10, 12-16, 19-21, 23-25, 42, 49, 52, 76, 95, 176
    irradiance, 95, 100
Erwinia, 133
Esterase, 115
Ethionine, 45
Ethylene, 100, 102, 189, 190
Exine, 115, 116, 118

Female sterility, 175
5-fluorodeoxyuridine, 47
5-fluorouracil, 45
Florigen (see flower stimulus)
Flower
    anthesis, 96
    determination, 52-56, 191, 194
    development, 10, 48, 95-97, 102, 103, 106, 117, 181, 190, 192, 194, 199
        in vitro, 117, 118, 181
    evocation, 1, 10, 20, 22, 42, 44, 45, 47, 48
    hormone, 10, 13
    induction, 1, 3, 6, 10, 19-24, 34, 42, 43, 45, 55, 79, 180, 182, 183, 195
    inhibitor, 10, 12, 34, 45, 48
    initiation, 10-13, 15, 19, 34, 42, 45, 47, 51-53, 55, 56, 76-78
    meristem, 42, 43-45, 48, 52, 53, 58, 76, 191, 196
    morphology, 48, 55, 177
    mutants, 10-16, 174, 177, 180, 181, 188
    proteins, 45, 183
    stimulus, 10, 16, 20, 42-44, 176

Genetic mosaic, 177
Gibberellin, 11, 29-40, 99-102, 115, 117-119, 177
    GA_1, 33, 34, 36-40, 100
    GA_3, 29, 30-40, 117
    GA_4, 29, 30, 33, 34, 38
    $GA_{4/7}$, 29, 31-33, 39, 99
    GA_7, 29, 33, 34, 39
    GA_8, 38
    GA_9, 29, 30, 38
    GA_{12}, 38
    GA_{20}, 38
    GA_{32}, 37-39
    synthesis mutant, 11
Glycine max, 175
Glycoprotein, 136, 140-143, 147-154, 156-159, 161, 162, 167, 168
Glycosylation, 136, 140, 161, 168, 170
GUS (β-glucuronidase), 130-132, 196
Gynoecium, 88, 106, 179, 192, 200

Hemerocallis flava, 179
Heterochrony, 190-192
Homeosis, 180, 192, 199
Hoop reinforcement, 60-71
Hordeum vulgare, 2, 3

Iberis, 146
Incompatibility genes
    Mod1, 152, 153
    SLG, 147, 153, 158-162
    SLR1, 158-162
    Sup1, 152, 153
    Sup2, 151-153

Inflorescence, 52, 95-103, 158
    development, 66-68, 79, 174, 181
Inhibitors
    blasticydin S, 140
    cycloheximide, 140, 143, 190
    floral, 10, 12, 45, 47, 48
    glycosylation, 142
    protein synthesis, 142
    puromycin, 140
Inositol, 7
In situ hybridization, 112, 128-130, 157-159
Ipomoea purpurea, 21, 22

Kalanchoe, 60, 61, 64, 65, 67, 68, 72
Kinematic analysis, 82, 200

Latex beads, 179
Leaf mRNA, 134, 176
Leaky gene, 12, 13, 153
Lemna, 5
Lesquerella, 146
Ligule, 177, 178
Lilium longiflorum, 82-93, 143, 178, 200
Lolium temulentum, 3, 23, 36-40, 45
Lupinus, 190
Lycopersicon esculentum, 95, 114, 134
Maize (see corn)
Male sterility, 114, 117, 118, 121, 128, 130, 134, 187
    cytoplasmic, 114, 118, 119, 121-125
        ogu-cms, 117, 119
        CMS-C, 122
        CMS-S, 121, 122
        CMS-T, 121-125
        Pcf, 121
    nuclear (genic), 114, 175
        GMS S1-2/S1-2, 118, 119
    nuclear restorer genes
        Rf, 1-4, 122
Megaspore, 118. 175
Meiosis, 100, 102, 108, 110, 111, 157, 175, 178, 187, 190
    proteins, 178
Meristem (see Flower)

Methomyl, 123
Microspore, 108, 110, 111, 115, 116-119, 128, 175, 178, 179, 187, 188
Microsporogenesis, 102, 114, 115, 118, 119, 121, 124, 128
Microtubule, 61
Mitochondria, 43, 44, 118, 121-125
    genes, 121-125, 168, 185
Modifier gene, 156
Morning glory, 195

Nicotiana, 54, 143, 146, 157
    alata, 163, 168-171, 184
    sylvestris, 52, 76, 79, 194
    tabacum, 52, 54, 76-79, 181, 191
        cv Maryland Mammoth, 15, 52, 53, 76, 77, 79, 191
        cv Wisc. 38, 52, 53, 58, 181, 191, 194

Oenothera biennis, 197
    organensis, 186
Ovary, 99, 118, 166, 179, 200

Papaver, 143
Passiflora caerulea, 197
Pea, 15
    flowering genes
        det, 14-16
        dm, 14, 16
        Dne, 10-16
        E, 11-16
        gigas, 15, 16
        Hr, 11-16
        Leaky gene, 12, 13
        Lf, 10-16
        lw, 11
        Sn, 10-16
        veg, 10, 11, 15
Pectate lyase, 132
Pellicle, 137, 138
Perilla, 23
Petals, 69, 70, 99, 121, 128, 180, 183, 192
Petunia, 121, 122, 134, 184, 190
Pharbitis nil, 6, 19-26, 34, 35, 38, 180, 195

Philodendron selloum, 197
Photoperiod (see environmental control), 1-3, 10, 12-16
Phytochrome, 1- 3, 5-7
Picea abies, 30
Pinaceae, 32
Pinus radiata, 32, 33, 39
Pistil, 108, 128, 136, 142, 157, 166, 167
Pisum (see Pea), 10, 11
Plant growth regulator (PGR), 96, 98-101, 103, 114,
Plastochron, 47, 48
Pollen, 108-112, 128-130, 134, 136-143, 146, 150, 156, 157, 185-188, 190, 191
    adhesion, 141, 156
    compatibility, 136, 193
    development, 108-111, 114, 115, 119, 125, 128, 130, 134, 136, 186, 187
    fertility, 122, 131
    generative cell, 108, 112, 134
    germination, 109-111, 129, 131, 140, 166
    hydration, 134, 137-142, 156, 166
    male gametophyte, 108
    sperm cells, 108, 112, 166
    tube growth, 108, 109, 111, 133, 140-142, 150, 156, 159, 167, 171, 179, 193
    vegetative cell, 108, 112, 134
    wall proteins, 134, 170, 179
Pollination, 137-142, 147-150, 156, 165, 189, 190
Polyamines, 20, 143
Promoter analysis, 128
Protein, 45, 134, 140, 141, 143, 148, 156, 168, 170, 179, 180, 183, 185, 189
    kinases, 7
Pseudotsuga menziesii, 29
Puromycin, 140, 141

Radish, 146
Raphanus raphanistrum, 146, 179

sativus, 117, 146, 193
Red clover (see *Trifolium pratense*), 185
Relative growth rate, 85, 200
    relative elemental growth rate, 86, 87
Restriction length fragment polymorphism, 107, 147-152, 159, 168
Roots, 52, 78, 79, 98, 111, 130, 134

Salicylic acid, 197
S-alleles, 147, 148, 150, 151, 184
Self-incompatibility, 146, 147, 156, 157, 165, 193
    gametophytic, 142, 166, 185
    sporophytic, 136, 137, 142, 157, 167
Sepals, 45, 68, 69, 99, 106, 180, 183, 192
Signal peptide, 159, 161
Silene coeli-rosa, 45
Sinapsis alba, 42-44, 46-49, 176
S-gene, 142, 161, 162, 165, 166, 170
S-locus, 146, 147, 151, 156, 184, 193
S-locus specific glycoprotein (SLSG), 143, 147-159, 162, 167, 185
S-multigene family, 158, 163
Soft rot disease, 133
Southern corn leaf blight, 122
Soybean, 175
Spinach, 44
Sporogenous tissue, 97, 100, 103, 130, 190
Sporopollenin, 187
Stamen, 70, 74, 87, 88, 99, 100, 102, 114, 117-119, 177, 180, 181
Stigma, 118, 121, 136-138, 140, 142, 146, 150, 151, 156-158, 161, 166, 190, 200
    glycoproteins, 140, 141, 143, 148, 156, 170
    papillae, 139, 140, 156
STS, 100, 190
Style, 108, 136, 166, 190, 200
    proteins, 168, 170, 185
    secretory matrix, 179
    transmitting tract, 179
Succinic dehydrogenase, 43
Suppressor gene, 146

Surface growth analysis, 83, 91, 200

Tapetum, 115, 118, 119, 124, 125, 187
Temperature sensitive mutant, 106
Tepal, 83-89, 93, 199
Thermogenecity, 197
Tobacco (see *Nicotiana*), 7, 15, 44, 53, 93, 130, 131, 134, 181, 191, 196
    CMS, 121
    shoot culture, 181
    thin cell layer, 76-78
Tomato (see *Lycopersicon*), 95-97, 99, 100, 102, 103, 114, 117, 128, 130, 131, 180, 182
    anantha mutant, 182
    ga-2 mutant, 100
    green pistillate mutant, 180
    stamenless-2 mutant, 100, 114
Tradescantia paludosa, 108-110
    virginiana, 197
Transgenic plant, 24, 128, 130, 134
    tobacco, 7, 131, 196
    tomato, 130, 131
Transposable element, 21, 26
Trifolium pratense, 185
Triticum aestivum, 187
Tryphine, 137
T-toxin, 123
Tunicamycin, 142, 170
T-urf13 [urf 13], 122-125

Vicia faba, 179
Vinca, 61, 65, 68

Waveform, 90, 91, 93
Wheat (see *Triticum aestivum*), 187

Xanthium, 23, 44

Zea mays, (see corn)